The Concept of Model

TRANSMISSION

Transmission denotes the transfer of information, objects or forces from one place to another, from one person to another. Transmission implies urgency, even emergency: a line humming, an alarm sounding, a messenger bearing news. Through Transmission interventions are supported, and opinions overturned. Transmission republishes classic works in philosophy, as it publishes works that re-examine classical philosophical thought. Transmission is the name for what takes place.

The Concept of Model

An Introduction to the Materialist Epistemology of Mathematics

Alain Badiou

Edited and translated by
Zachary Luke Fraser and Tzuchien Tho

re.press

PO Box 75, Seddon, 3011, Melbourne, Australia
http://www.re-press.org

British Library Cataloguing-in-Publication Data
A catalogue record for this book is available from the British Library

Library of Congress Cataloguing-in-Publication Data
A catalogue record for this book is available from the Library of Congress

National Library of Australia Cataloguing-in-Publication Data

Badiou, Alain.
[Le Concept de modèle. English.]
The concept of model : an introduction to the materialist
epistemology of mathematics.

Bibliography.
ISBN 9780980305234 (pbk.).

1. Model theory. 2. Mathematics - Philosophy. I. Title.
(Series : Transmission).

510

Designed and Typeset by *A&R*

Printed on-demand in Australia, the United Kingdom and the United States
This book is produced sustainably using plantation timber, and printed in the destination market on demand reducing wastage and excess transport.

Contents

Acknowledgements

First of all, I thank Shan Mackenzie-Fraser, my wife and friend; it's far too little to say that your encouragement, patience and love made this project possible, but it's a start. Thanks, too, to my brilliant children, Ruis, Kai, Hinahni and Sophie; someday, I'll tell you all about the difficult marriage of formal semantics and the class struggle that kept your father busy back in the summer of 2007. My gratitude also goes to my parents, Zach and Marion Fraser, whose enduring support and confidence, babysitting shifts and ribbings, helped me get this book to press. I also wish to thank my comrade and collaborator, Tzuchien Tho, who, above and beyond the achievements explicitly registered in this book, spent countless hours running around Hell's creation to track down obscure documents, transcribe and translate long-forgotten videotapes, and wait repeatedly for appointments with Beckettian personages in Kafkaesque environs. I am grateful, as well, to the numerous men and women who assisted in reading torturously awkward drafts of this translation: Rohit Dalvi, who, some years back, suggested that I write a master's thesis on Badiou's thought, and who, when that project tangentially spawned a rough translation of the *Le Concept de modèle* (which at the time was little more than an exercise and a hobbyhorse), surprised me by saying that it wouldn't be inconceivable to publish the thing; Gillman Payette and Stephanie Dick, who made up two thirds of a short lived philosophy of mathematics reading group that dedicated more than half its meetings to *The Concept of Model*; and Gordon McOuat, who offered consistently helpful feedback on early drafts of this translation, despite his distaste for all things Althusserian. I would also like to thank the aforementioned circle of translators and scholars that helped bring this text to fruition, particularly: Ray Brassier, Oliver Feltham, Robin Mackay, and Alberto Toscano. Thanks also to Adam Bartlett, Paul Ashton and Justin Clemens at re.press, whose camaraderie and commitment have made this entire project a pleasure to be a part of, and whose fidelity to ideals higher than the marketplace has made it an honour. Finally, thanks to Alain Badiou, both for giving my questions and critiques the dignity worthy of problems demanding solutions, and, of course, for writing.

Zachary Luke Fraser

Conditions for the possibility of philosophical work are always split between the alimentation of daily life and the clinamatic impact of friendship. Both are aleatory by nature and their ranking cannot but be arbitrary. Zachary Luke Fraser extended a hand to me in camaraderie and partnership; what he sometimes received was a fist full of criticism and immature scribbling. I owe him gratitude and apology in equal amounts. Amy Anderson's evolving praxis of our shared fidelity to the amorous truth condition has pushed me to continuously re-examine the indiscernible in being. The last year would not have been possible without her and Mr. Marx's continuous encouragement. I thank Mr. Tu Chiang-shiang for a love of books and Mr. Tu Nai-ping for teaching me the undesirability of a merely nutritive life. Madame Liu Yue-chiao has always been a stronghold and never asks anything in return. My parents, Tho Kee-ping and Tu Hsiu-chen, taught me how to 'zuo-ren', roughly translated as 'be human', a gift that can neither be repaid nor returned. David and Karen Anderson have shown me a degree of grace that I have not experienced before. I thank Bruno Besana, Oliver Feltham, François Farellacci, Colin Mcquillan, Nate Holdren, Brian Howell for camaraderie and friendship, and O. Bradley Bassler for late night clarifications on incompleteness. Alain Badiou can be blamed for the number of youths corrupted by his philosophical practice. Forty years after *Le Concept de modèle*, I consider myself lucky to have been perverted.

Tzuchien Tho

Introduction

The Category of Formalization: From Epistemological Break to Truth Procedure

Zachary Luke Fraser

Really, in the end, I have only one question: what is the new in a situation? My unique philosophical question, I would say, is the following: can we think that there is something new in the situation, not outside the situation nor the new somewhere else, but can we really think through novelty and treat it in the situation? The system of philosophical answers that I elaborate, whatever its complexity may be, is subordinated to that question and to no other. [...] But, of course, to think the new in situation, we also have to think the situation, and thus we have to think what is repetition, what is the old, what is not new, and after that we have to think the new.

At least in this regard I remain more profoundly Hegelian. That is, I am convinced that the new can only be thought as process. There certainly is novelty in the event's upsurge, but this novelty is always evanescent. That is not where we can pinpoint the new in its materiality. But that is precisely the point that interests me: the materiality of the new.

— *Alain Badiou, in interview with Bruno Bosteels*[1]

The author of a work always likes to say that they have constantly evolved but I would like to pose the idea of Bergson that a philosopher only has one idea. If we suppose only one idea, it is this idea. I believe that if all creative thought is in reality the invention of a new mode of formalization, then that thought is the invention of a form. Thus if every creative thought is the invention of a new form, then it will also bring new possibilities of asking, in the end, 'what is a form?' If this is true, then one should investigate the resources for this. As a resource, there is nothing deeper than that which the particularity of mathematics has to offer. This is what I think, I held this point of view and I hold it now. It is not that mathematics is the most important, not at all. Mathematics is very particular but in this, philosophically speaking, there is something that is specifically tied to mathematics in the very place of thought. [...] Mathematics holds something of the secret of thinking.

— *Alain Badiou, in interview with Tzuchien Tho*[2]

---

1. 'Can Change be Thought', in Gabriel Riera (ed.), *Alain Badiou: Philosophy and its Conditions*, Albany, SUNY, 2005, pp.252-3.

2. See pp. 102-3 below.

The Concept of Model, Alain Badiou's first book of philosophy, is both concise and explicit, and there's no need to burden it with summary.[3] Its goals, its theses, its arguments and examples are drawn out with admirable didactic clarity. To rough them out in advance and threaten the text with unnecessary doubling, would only obfuscate things more than it would clarify them. If there is obscurity in this text, it is legible only in retrospect, through the conceptual machinery developed in the almost forty years that followed its initial (interrupted) delivery in Louis Althusser's *Philosophy Course for Scientists.* That is to say, what is not initially clear is how this work participates in Badiou's general philosophical project. It is this that needs explaining.

Of course, it is first necessary to make out what, precisely, this project *is*. Taking up Badiou's own suggestions, reprinted above, I will say this: it is a project aspiring to think the situated and material emergence of novelty, and the universality by which it subtracts itself from the old, by means of the fundamental idea of *formalization*. This project begins, in the late nineteen-sixties, with a series of preliminary meditations on the resources harboured by mathematical logic for this specifically philosophical task. It is amidst these early studies that *Le Concept de modèle* was written, under conditions I will address in the next section. It is from this text, translated here, that I will take my bearings, but my concern, in the end, will be the trajectory that the category of formalization carves out through the course of Badiou's thought, determining its notion of the new. This will take us, roughly speaking, from the epistemological writings of the late sixties to the lessons recently published in *The Century*.

I. Ideology and Epistemological Break

In order to understand how mathematics in general, and model theory in particular, come to play such a central role in Badiou's early theorization of novelty, we should begin by outlining the general conception of novelty within which these early investigations take place. At the time when Badiou first broaches the question of novelty, the context in which he is working is a faithfully Althusserian one. This much is evident on the very surface of Badiou's early works, whose basic fidelity to the Althusserian project is (sometimes implicitly, but often explicitly) declared on every available occasion between 1966 and 1969. The particular texts I have in

3. Ray Brassier has, in fact, already provided a fine exegesis of the text. See 'On Badiou's Materialist Epistemology of Mathematics', *Angelaki*, vol. 10, no. 2, 2005, pp. 135-150.

mind are: 1966's 'Le (Re)commencement du matérialisme dialectique' ('The (Re)commencement of Dialectical Materialism'), 1968's 'La subversion infinitesimale' ('The Infinitesimal Subversion'), 1969's 'Marque et manque: à propos de zéro' ('Mark and Lack: On Zero') (which was, in fact, penned in 1967), and, of course, *The Concept of Model*, which, as an instalment in the *Philosophy Course for Scientists*, had the most immediate connection of Badiou's early texts to the Althusserian undertaking. The date of these lessons is particularly significant: as the book's original foreword informs us, the first lesson was delivered on 29 April 1968, and contained the material that would make up the first five chapters of the present book. The remaining five chapters, we are told,

> were to be the object of a second lecture, due to be given on May 13th, 1968. That day, as is well known, the popular masses, mobilizing against the bourgeois, Gaullist dictatorship, affirmed their determination across the entire country, and began the process that would lead to a far-reaching confrontation between the classes, turning the political conjuncture on its head and provoking effects whose aftermath was not long in coming.

These events can accurately be said to mark the end of Badiou's Althusserian period, and, as can be gathered from the foreword, consigned the project that they 'happily interrupted' to a 'bygone conjuncture', such that a mere six months after the first lesson was given, they would be presented more as an historical artefact than a contemporary intervention. Althusser, who, for medical reasons, did not participate in the events of May in person, would make some minor modifications to his general project in the years that followed; the resulting '*autocritique*' was drawn up in 1974 in his 'Elements of Self-Criticism'.[4] The events of May cut far more deeply into Badiou's trajectory than they did into Althusser's, however, and in 1975, he would retrospectively date his break from the Althusserian project, to the Paris uprising that would be for him 'a veritable road to Damascus'.[5]

From that moment forward, the celebrated notion of the 'event' would increasingly come to occupy Badiou's thought, though the idea in its present (and still developing) form would take some time to

4. Louis Althusser, *Eléments d'autocritique*, Paris, Librairie Hachette, 1974. Translated as 'Elements of Self-Criticism', in *Essays in Self-Criticism*, trans. Grahame Lock, London, New Left Books, 1976.

5. Alain Badiou, *Théorie de la contradiction*, Paris, Maspero, 1975.

crystallize. It is not with the 'event' that Badiou's preoccupation with the emergence of novelty begins however, nor, even now, is it the most central or significant category at work in this capacity.[6] At the risk of subordinating the (by no means trivial) rupture that the Events of May mark in Badiou's philosophy, to what, of the Althusserian project, continues uninterrupted into his later thought, I would argue that the figure of the event is altogether secondary to the procedural conception of novelty that Badiou inherits from his teacher. Badiou's invocation of 'events', over which a disproportionate amount of ink has already been spilled, is nothing but a fascinating and mystifying snare if it is left without a proper understanding of the arduous and protracted *procedures* through which the new pulls itself away from the old. We should remember Badiou's remark, in Mediation Twenty of *Being and Event*, that 'the event is only possible if special procedures conserve the evental nature of its consequences. This is why its sole foundation lies in a *discipline* of time...'[7] That, in the full-fledged metaphysical framework of *Being and Event*, truth procedures follow *after* the events to which they pledge their fealty, does not alter the *conceptual priority* of the procedure itself, a priority that this passage serves to underscore. What is 'evental' in the event—the novelty it heralds—is largely external to the event itself; it is a production that can only come afterwards. This conceptual priority is coupled with philological precedence: an effort to unfold the general structure of such 'procedures' can already be detected in Badiou's earliest writings, beginning with the 'pre-evental' set to which the current book belongs. I will therefore set

6. Badiou's remarks in the already-quoted interview with Bosteels are instructive in this respect:

> the principle contribution of my work does not consist in opposing the situation to the event. In a certain sense, that is something everybody does these days. The principle contribution consists in posing the following question: what can be deduced, or inferred, from there, from the point of view of the situation itself? Ultimately it is the situation that interests me.
>
> I don't think that we can grasp completely what a trajectory of truth is in a situation without the hypothesis of the absolute, or radical, arrival of an event. Okay. But in the end, what interests me is the situational unfolding of the event, and not the transcendence or entrenchment of the event itself. Thus, in my eyes, the fundamental categories are those of *genericity* and of *forcing*. Genericity can be understood as the trajectory of aleatory consequences, which are all suspended from whatever the trace of the event is in the situation; and forcing consists in the equally extremely complex and hypothetical way in which truths, including political truths, influence and displace the general system of our encyclopedias, and thus, of knowledge. Badiou and Bosteels, 'Can Change be Thought?', p. 252

These are the concepts whose genealogies we will attempt to trace.

7. Alain Badiou, *Being and Event*, trans. Oliver Feltham, London, Continuum, 2005, p. 211. Henceforth cited BE.

the notion of the event aside entirely for the time being, in order to explicate that through which Badiou sought to articulate the new and the universal prior to the theorization of 'the event' that the Events of May provoked.

What most preoccupied Badiou within the context of Althusserian epistemology was the category of the *epistemological break*, the category through which Althusser and his circle sought to understand the separation of the scientific from the ideological. The term itself is somewhat unfortunate, and misleading in its connotations of suddenness. It tempts us to imagine the break as a specific *instant*, a singular historical moment. It tempts us, moreover, to draw premature analogies between epistemological break and event. For both Althusser, as well as for his teacher Bachelard, from whom he inherited the term,[8] what is at issue in the epistemological break is not an instant in time but an ongoing process, an interminable struggle between the scientific and the ideological at the heart of scientific practice. For Bachelard, the break, or 'epistemological rupture' as he called it, realized the triumph of scientific conceptualization over the imaginary and cognitive habits that obstruct its path—these, he called 'epistemological obstacles'. Such obstacles, according to Bachelard, cling to thought like its own shadow, and require constant and vigilant correction. 'It is at the very heart of the act of cognition', he writes in *The Formation of the Scientific Mind*, 'that, by some sluggishness and disturbances arise. [...] Knowledge of reality is a light that always casts a shadow in some nook or cranny. It is never immediate, never complete. Revelations of reality are always recurrent'.[9] It is only through the recurrent overturning of and rupture with what it thinks it knows, that science makes progress. Althusser incorporates the idea of the break into a Marxist (though rather idiosyncratic) theory of ideology, under which the motley set of 'epistemological obstacles' cohere into a vast ideological

8. A slight difference exists between Bachelard and Althusser's terminology, which generally does not survive their translations into English: where Bachelard has '*rupture epistemologique*', Althusser and Badiou have '*coupure epistemologique*'. This slight difference is explained as follows, in Althusser's 'Elements of Self-Criticism': 'Every recognized science not only has emerged from its own prehistory, but continues endlessly to do so (its prehistory remains always contemporary: something like its *Alter Ego*) by *rejecting* what it considers to be *error*, according to the process which Bachelard called "the epistemological *break* [*rupture*]". I owe this idea to him, but to give it (to use a metaphor) the sharpest possible cutting-edge, I called it the "epistemological break [*coupure*]". And I made it the central category of my first essays', 'Elements of Self-Criticism', p. 114.

9. Gaston Bachelard, *The Formation of the Scientific Mind*, trans. Mary MacAllester Jones, London, Clinamen Press, 2002, p.24.

complex (at the price, perhaps, of the psychoanalytic subtlety and diversity of Bachelard's researches). The break, for Althusser, as for his student, Badiou, is understood as a break *with ideology*, or, to be more precise and more Bachelardian, with the ideological immanent to science itself.

In 'Le (Re)commencement du matérialisme dialectique', a 1966 review of Althusser's *Pour Marx, Lire Capital* and 'Matérialisme historique et matérialisme dialectique', Badiou gives a general exposition of Althusser's theory of ideology, marked by an idiosyncratic series of inflections that will become a lasting motif through his work—one which will resurface whenever the broad notion of the 'status quo' is at stake. In this text, 'ideology' decomposes into a threefold function of *repetition*, *totalization* and *placement*. Ideology expresses these functions by: (a) instituting the repetition of immediate givens in a 'system of representations [...] thereby produc[ing] an effect of *recognition* [*reconnaissance*] rather than cognition [*connaissance*]'[10] (RMD 449); (b) charging this repetitional system with a unifying sense of worldhood and totality while ordaining it as 'Truth';[11] (c) reinscribing both individuals (as subjects commanded to 'take their place') and scientific concepts in this representational whole.[12]

Of course, the ultimate function of ideology is, as Badiou notes parenthetically, 'to serve the needs [*les besoins*] of a class' (RMD 451, n.19). It is this dimension of ideology that is, by all rights, most crucial to any Marxist philosophy that maintains the communist project in view. 'The (Re)commencement' has little to say of the exact relation between ideology and the class struggle—it mentions in passing that some such relation exists and ends it at that—but in this respect the text is a legitimate offspring of Althusser's own work. Several years later, in his 'Elements of Self-Criticism', Althusser would diagnose his tendency to dissociate ideology and the class struggle as a symptom of his work's 'theoreticist deviation', which expressed itself in a habitual conflation of the opposition between the classes with distinction between science and ideology, a distinction that, itself, is mapped too hastily onto the schism between

10. Alain Badiou, 'Le (re)commencement du matérialisme dialectique', in *Critique*, vol. 23, no. 240, May 1967, p. 449. Henceforth cited RMD.

11. 'The connected system of designators re-produces the unity of existences in a normative complex that *legitimates* the phenomenal given (what Marx calls appearance). As Althusser says, ideology produces the *feeling* of the theoretical. The imaginary thus announces itself in the relation to the 'world' as a *unifying pressure*, and the function of the global system is to furnish a legitimating conception of *all* that is given as real' (RMD 450-1).

12. Ideology is 'a practico-social *function* that commands the subject to "take its place"' (RMD 450). And, it 're-inscribes in the re-presented immediacy the very concepts of science itself' (RMD 450, n.19).

truth and error.[13] Ideology, seen through the bias of this 'theoreticist deviation', is quickly reduced to a monolithic system of illusion, which offers nothing to science but errors. Several years after the publication of the present book, Badiou would collaborate with François Balmès to produce a devastating attack on the Althusserian theory of ideology in a short book entitled, *De l'idéologie* (*Of Ideology*). Their essay is, above all, a critique of the monolithic and illusional nature that Althusser attributes to ideology;[14] it sought, against Althusser, to lay bare:

- the relation between ideology and the reality of historical phenomena of exploitation and oppression;
- the divided, conflictual character of the ideological sphere. Or, in other words, the necessary subordination of the definition of ideology to the reality of the ideological struggle.[15]

Both themes are already intimated in *The Concept of Model*, whatever confessions of 'theoreticism' may be attached to the text, though they are, to be sure, fenced within the narrow domain of logico-mathematical

---

13. At the beginning of his 'self-criticism', Althusser explains his mistake:

> I wanted to defend Marxism against the real dangers of *bourgeois* ideology: it was necessary to stress its revolutionary new character; it was therefore necessary to 'prove' that there is an antagonism between Marxism and bourgeois ideology, that Marxism could not have developed in Marx or in the labour movement except given a radical and unremitting *break* with bourgeois ideology, an unceasing struggle against the assaults of this ideology. This thesis was correct. It still is correct.
>
> But instead of explaining this *historical* fact in all its dimensions—social, political, ideological and theoretical—I reduced it to a simple *theoretical* fact: to the epistemological '*break*' which can be observed in Marx's works from 1845 onwards. As a consequence I was led to give a *rationalist* explanation of the 'break', contrasting *truth* and *error* in the form of the speculative distinction between *science* and *ideology*, in the singular and in general. The contrast between Marxism and bourgeois ideology thus became simply a special case of this distinction. Reduction + interpretation: from this rationalist-speculative drama, the class struggle was practically absent.
>
> All the effects of my theoreticism derive from this rationalist-speculative interpretation. Althusser, 'Elements of Self-Criticism', pp. 105-6.

14. Perhaps the most damning objection that Balmès and Badiou make against Althusser is that the monolithic character of his conception of ideology is a consequence of his stubbornly bourgeois class position. On pages 36-7 of *De l'idéologie*, we read: 'We can clearly see where the difficulty is for Althusser: to seize ideologies as processes of scission demands the point of view of a particular class: in effect, *it's the point of view of the oppressed classes that yields the experience of divided ideology*. The dominant class practices and imposes its own ideology as the dominant one, presenting it as unique and unifying. It's the dominated classes that reveal the mystification of the unifying ideology, on the basis of practices of classes in revolt, *irrepresentable in the dominant ideology*. A project in the general theory of ideology that does not inscribe this division in the very essence of the phenomenon warrants our suspicion that it has not taken the point of view of the oppressed'.

15. Alain Badiou, *De l'idéologie*, Paris, Maspero, 1976, p. 27.

practice and associate with the larger political struggle only by way of a few, rather vague, allusions made in the book's final chapter. What is of most immediate interest is Badiou's shift from the Althusserian identification of ideology and error to a far more general fusion between ideology as a whole and what, according to our list, is its first attribute: representation.

This fusion is less an independent thesis in Badiou's early work than a predominant tendency. Already in his earliest texts, the representational dimension of ideology significantly outweighs the others, and suffers the greater part of Badiou's critical attention. The emphasis placed on this particular attribute is so strong that it quickly bleeds over into a suspicion directed towards representation *as such*. Repeatedly in Badiou's text, we find indications of a slide from a position that sees ideology as partaking in various operations of representation (it is '*a* process of repetition', which 'produces *an* effect of recognition' (RMD 449)), to a wholesale subsumption of the representational by the ideological (at one point the general forms of 'reproductive discourse' are directly equated with 'the most abstract forms of any ideological discourse' (RMD 450, n.17). More than anything else, this move towards implicitly associating representation *as such* with the ideological plane calls to mind the hyperbolic doubt. In Cartesian fashion, Badiou proceeds by casting aside all that is in the least reducible to designation or representation, no less than if he were to discover it to be absolutely ideological, and by these means seek to isolate what is essentially, because irreducibly, scientific.[16]

The conceptual metastasis of the notion of ideology throughout its primary attribute, while, perhaps, setting ideology at some distance from the concrete struggles that animate it, has the effect of prohibiting any clear-cut separation of the scientific from the ideological akin to the schism between truth and error (a schism which, it must be said, is less evident in Althusser's own early writings than his self-criticism leads one to believe—we should keep in mind, after all, that Badiou sees himself as being essentially *faithful* to Althusser in these early texts). If representation as such is always at least potentially ideological, there can be no question of *purifying* science of ideology. While it is the wager of Badiou's entire epistemological effort that science cannot be *reduced* to

16. See René Descartes, *Meditations on First Philosophy*, trans. John Veitch, London, Everyman's Library, 1969, Mediation II, p. 85: 'I will [...] proceed by casting aside all that admits of the slightest doubt, not less than if I had discovered it to be absolutely false; and I will continue always in this track until I shall find something that is certain, or at least, if I can do nothing more, until I shall know with certainty that there is nothing certain'.

representational activity, it is nonetheless impossible to say, without offending every experience of scientific practice, that science has *nothing to do* with representation. The proposition that science has an inalienable need of representational operations, in fact, serves Badiou as the essential bridge between science and ideology, for

> save for repeating that science is science of its object, which is a pure tautology, the question, 'Of what is science, science?' admits of no other answer than: science produces knowledge of an object whose *existence is designated* by a determinate ideological region. (RMD, 450)

Or, to put it another way, 'the material of science is, *in the last instance*, ideology'.[17] The utter dependency of science on ideology for the sheer manifestation of its initial objects, and for the representational screen by means of which it can 'see' what it is doing, prohibits us from attempting any sort of surgical excision of the ideological from the scientific. The categorical opposition between science and ideology, as necessary as it may be for epistemology, 'does not permit any immediate classification of various practices and discourses; even less does it license the abstract "valorization" of science "against" ideology' (RMD 450). The question that confronts us, and which, for the Badiou of 1966, stands out as the 'apex of dialectical materialism', is not, 'How can science be purged of ideology?' but rather: 'How can we think the articulation of science onto what it is not, all the while preserving the impure radicality of the difference?' (RMD 452). Insofar as the production of knowledge is ultimately irreducible to the ideological terrain on which it takes place, it is distinctly non-representational. Indeed, the difference between science and ideology is the difference between a 'process of transformation' and a 'process of repetition'; the work of science is not primarily effected through representation, but 'through the ruled production of an object essentially distinct from the object that is given—distinct, even, from the real object' (RMD 449).

'The epistemological break' is the name given to the impure but intractable articulation of processes of material and conceptual transformation that the sciences effectuate in the midst of the ideological 'given'. It is the focal point of the Althusserians' implicit definition of science, which can be formulated: *Science is science insofar as, within ideology, it perpetuates an (impure but persistent) 'epistemological break'.* This definition holds

17. Alain Badiou, 'Marque et manque: à propos de zéro' in *Cahiers pour l'analyse*, vol. 10, p.165. Henceforth cited MM.

true even of the 'a priori' science of mathematics, whose

> singularity lies in the fact that its determinate 'exterior' is nothing other than the ideological region wherein *mathematics designates itself*. Such is the real content of the 'a priori' character of that science: it breaks only with its *own activity* [*fait*] as designated in its own re-presentation ... (RMD 453, n.21)

The thesis articulated here is one that endures throughout Badiou's early work, and is repeated almost verbatim in 1967's '*Marque et manque: à propos de zéro*' where he 'define[s] an "a priori" science as one which has business with ideology only insofar as it is represented by the latter: science breaks incessantly with its own representation in the re-presentational space' (MM 165).

Now, one way in which mathematics produces a representation of its own activity is the logico-mathematical theory of models. There's no point belabouring the details of the theory here: they're taken up at length in Badiou's own text, and explained there with remarkable clarity. For now, let us oversimplify somewhat and say that what, in general, model theory constructs is a logical analysis of mathematics' own 're-presentational space', weaving together meticulously ruled systems of interpretation between formal systems and the various mathematical structures that those systems can be said to be 'describing'. Owing to the precision and generality of the mathematical analysis of 'representation' that model theory produces, it was not long before the discipline attracted the attention of epistemologists, particularly those of a logical-empiricist bent, who were quick to see in the syntax and semantics of the emerging science a sort of mathematical analogue of their own central dichotomy of logic and observation. Formal semantics seemed to provide a concept of representation whose kinship with the notion germinating in the logical empiricism of the early twentieth century was undeniable. This kinship seemed to signal the breathtaking possibility of a science of science—a scientific study of science-as-representation.[18] It was not long before the

18. The characterization of logical empiricism as a *primarily* representationalist account of science is Badiou's. I am not yet convinced that this characterization is either accurate on the whole or attentive to the peculiarities of logical empiricism. For my part, I am inclined to argue that the simplistic identification of science as the activity of producing accurate representations of the world has more in common with Lenin than any of the positivists! It was Lenin, after all, who, in *Materialism and Empirio-criticism*, condemned as idealist sophistry any doctrine failing to recognize that 'the "objective truth" of thinking means *nothing else* than the *existence* of objects (i.e., "things-in-themselves") *truly* reflected by thinking!', V. I. Lenin, *Materialism and*

explicit synthesis of the two traditions came into effect, most notably in the work of such masterful philosopher-logicians as Rudolf Carnap. It is this effort at synthesis on the part of logical empiricism that constitutes the primary target of Badiou's critique in *The Concept of Model*. We can find an important anticipation of this attack in a footnote to 'The (Re) commencement', where, following a remark that '[t]he notions of ideology can, in fact, be described as *designators*', Badiou writes:

> The formal theory of designation and, in general, formal semantics such as it has been developed by Anglo-Saxon logical empiricism, to my mind furnishes a structural analysis *of ideology*. Naturally, for Carnap, semantics is a theory *of science*: this is because logical empiricism is itself an ideology. It nevertheless undertakes the systematic sublation [*relevé*] of the general forms of *connected description* [*description liée*], of reproductive discourse, that is to say, of the most abstract forms of any ideological discourse. (RMD 450, n.17)[19]

What motivates this early attack on Carnap (and the peculiar, seemingly abandoned suggestion for the interpretation of semantics as an analysis of ideology) seems to be little more than the aforementioned identification of the representational and the ideological, and the accompanying worry that the thorough digestion of science by semantic analysis will result in an annulment of anything thinkable as an 'epistemological break'. In the

---

Empirio-criticism, trans. Anon., Peking, Foreign Languages Press, 1972, p. 112. The prime representatives of such sophistry? The early positivists! It is a small irony that Badiou prefaces the anti-representationalist manifesto found in *The Concept of Model* with an uncritical invocation of Lenin's book.

19. What Badiou commits to in this passage is not immediately transparent. To say that the general forms of reproductive discourse are 'the most abstract forms of any ideological discourse' is not to reduce ideology to representation, or to determine the latter as the essence of the former, but there are ambiguous tendencies in that direction. This ambiguity is in part due to the difficulties involved in the qualifier, 'most abstract'. Calling formal semantics an 'abstract' description of ideological discourse does not at all restrict it to describing ideological discourse *alone*, especially not if its intention is to provide an abstract description of *all* meaningful discourse! To call it an abstract description is only to say that it furnishes the *necessary*, though likely not the *sufficient* conditions for what it describes. Calling it the 'most abstract' description of ideological discourse, however, is a quite different move: abstraction, generally speaking, can proceed *indefinitely*, until the function of describing *anything* is altogether abjured. Following a certain scholastic line of reasoning, we might say that the 'most abstract' description we could offer of ideology (or anything else under the sun) would simply be 'that it *is*'. Implicit in the term 'most abstract description', if it is to make any sense at all, is the notion that it is the most abstract description *sufficient* for the isolation of what is being described. In this case, formal semantics is being painted as the most abstract description sufficient for the separation of the ideological from the scientific.

present book, Badiou significantly tempers his anti-empiricist critique: first, by rescinding the accusation that 'logical empiricism is itself an ideology'; it is, rather, a *philosophy*, which is to say, the 'ideological recovery' of science—the synthesis of ideological notions and scientific concepts into philosophical categories, to invoke the typology Badiou lays out at the beginning of the book. More importantly, and in accordance with the acknowledged impossibility of dealing with science and ideology as discrete and juxtaposable elements, Badiou displaces his critique from a simple *rejection* of the semantical picture of science towards more cautious investigation of how the semantical dimension of science—of mathematics, to be precise—can be understood according to the principles of a dialectical-materialist epistemology (adapted from Althusser's). To this end, Badiou insists that the semantical relation must be grasped as both *internal* to a given scientific—or, to be precise, mathematic—situation (rather than conceived as a relation, say, between theory and observation), and, insofar as it is still bound up with ideological representation, as the *site* of the epistemological break—the place where mathematics 'breaks incessantly with its own representation in the re-presentational space' (MM 165).

No longer the emblem of mere imitation, the concept of model is taken up as an integral element of the epistemological break. For the synthesis to hold, however, the idea of the latter must be formulated in such a way as to permit an answer to the fundamental question: 'How can we think the articulation of science onto what it is not, all the while preserving the impure radicality of the difference?' (RMD 452). Now, if semantics *is* the articulation of the scientific in mathematics onto the ideological, if it is the ideological reproduction of the scientific given, then the epistemological break must be that which emerges from and transforms the possibilities of this kind of articulation; it must be 'that by which a mathematical region, in taking its place as a model, finds itself *transformed*, tested, and experimented upon, as concerns the state of its rigour or generality': this, for Badiou, is precisely what takes place in the work of *formalization*, the production of the formal apparatuses of which the models are models. What follows is a far more complex picture of the relation between ideology and science, and between epistemological break and ideological representation, constructed around a properly philosophical *category* of model that differs markedly from both its positivist counterpart and the summary critique that Badiou previously cast its way:

> The category of model thus designates the retroactive causality

> of formalism on its own scientific history, the history conjoining object and use. And the history of formalism will be the anticipatory intelligibility of that which it retrospectively constitutes as its model (CM 54).
>
> The problem is not, and cannot be, that of the representational relations between the model and the concrete, or between the formal and the models. The problem is that of *the history of formalization*. 'Model' designates the network traversed by the retroactions and anticipations that weave this history: whether it be designated, in anticipation, as break, or in retrospect, as remaking (CM 54-5).[20]

The development that Badiou's epistemology undergoes between 'The (Re)commencement' and *The Concept of Model* is no doubt a significant advance in the sophistication and complexity of his philosophy. Several dubious oversimplifications in Badiou's previous intertwining of semantics and ideology seem to be cleared away by the Spring of '68, but as a consequence we are brought face to face with difficulties that are all the more complex. For instance: if the epistemological break, understood in terms of formalization, interiorizes the ideological representations from which it separates and reconfigures them as models, it is clear that the (wrongheaded and oversimplifying) opposition between syntax and semantics fails entirely in the schematization of the scientific and the ideological—what then, is capable of taking its place? If the break is to be thought as formalization, how can we obtain a scientific schema of ideology (if formalizing ideology is precisely what de-ideologizes it)? Clearly, we can proceed no further until we have better examined the concepts at our disposal.

II. Formalization: Subtraction and Forcing

Leaving room for subsequent modifications, we may call formalization any process by which a relatively informal practice or body of thought[21] (such as pre-Euclidean geometry) is organized into an axiomatic system, or, speaking in a more modern context, the process by which one brings

20. The important category of *remaking* [*refonte*] is one I must leave aside for the time being. A brief explanation of the term can be found in footnote 2 on p. 55 below.

21. I write 'relatively informal' because it is quite difficult to imagine what an absolutely informal 'body of thought' could possibly be, and, if, as Badiou argues in the interview in this volume, thought in its essence *is* the invention of form, then the expression, 'informal thought', if taken in a non-relative sense, would be a simple contradiction in terms.

a partially formalized system (such as Euclid's own) to a stricter form of rigour, unfolding the intuitive and definitional apparatus into an explicit axiomatic prescribing the ruled manipulation of a determinate set of bare symbols. In a monograph on the topic, R. Blanché offers a helpful outline of the norms of formalization that became dominant in the early twentieth century, by way of listing 'the fundamental conditions which a deductive presentation must satisfy if it is to be fully rigorous'. In this list, we have:

1. Explicit enumeration of the primitive terms for subsequent use in definitions.
2. Explicit enumeration of the primitive propositions for subsequent use in demonstrations.
3. The relations between the primitive terms shall be purely logical relations, independent of any concrete meaning which may be given to the terms.
4. These relations alone shall occur in the demonstrations, and independently of the meaning of the terms so related (this precludes in particular, relying in any way on diagrams).[22]

The nakedness of symbols and non-reliance on such intuitive apparatuses as diagrams and definitions[23] in a 'fully rigorous' deductive presentation are strictly correlative to the explicitness and sufficiency of its axiomatic, whose operation should have no need of intuitive notions lurking in the background nor content hiding in the notation.[24] The guiding idea is that, even though mathematics is compelled to operate on ideological terrain, and has no other points of departure than ideology's notional designators, it is within its power to achieve an epistemic independence, or indifference, with respect to this terrain. Such indifference is won through a laborious extraction of the terrain's obscure rationality, crystallizing it into a sequence of pure operational rules. In practice, this means producing a series of axioms, fixed in a strictly formulated symbolic language, in such a way that any problem that can be posed

22. R. Blanché, *Axiomatics*, trans. G.B. Keene, London, Routledge & Kegan Paul, 1962, pp. 21-22.

23. Definitions are not relied upon in such a system insofar as they are merely abbreviations. For example, it is entirely possible to eliminate the term Ø and its definition from set theory, and merely speak of 'some set *x* such that for all *y*, *y* does not belong to *x*'. It would only be more tedious.

24. It is nevertheless a psychological fact that, without an intuitive grasp of the system, one will have a great deal of trouble producing anything interesting within it. What is important is that the system, formally considered, is rigorous enough that intuition need not fill any *logical* gaps in its mechanism.

in the language can either (*a*) be solved solely by means of the resources disposed by the axioms, or (*b*) be revealed as *insoluble*, and as a 'weak link' at which a new formalizing rupture becomes possible.[25] Instances of the latter kind have been famously proven by Kurt Gödel to *necessarily* exist in formal systems of sufficient power (those capable of expressing elementary arithmetic).[26]

The elucidation of an ideologically represented domain into rigorously determined relations allows one to escape reliance on the recognizable content of the designational notions with which one began. A paradigmatic example of this procedure can be found in Hilbert's masterful transfiguration of Euclidean geometry, undertaken in *The Foundations of Geometry* of 1899. This text's most dramatic characteristic is its *utter absence of initial definitions*, that is, of all definitions that are not reducible to mere abbreviations. Indeed, if, with Kant, we accept that to define is to present 'the complete, original concept of a thing within the limits of its concept',[27] then we can take Hilbert to have shown that definition is not only unnecessary, but, as he compellingly argues in his famous debate with Frege, utterly impracticable in the field of mathematics. It is impracticable, Hilbert writes, because

> only the whole structure of axioms yields a complete definition. Every axiom contributes something to the definition, and hence every new axiom changes the concept. A 'point' in Euclidean, non-Euclidean, Archimedian and non-

25. Practically speaking, the requirement that *any* formally intelligible problem be soluble is a *heuristic impetus* guiding the creation of new axioms and not a criterion that measures axiomatic formalization's success. The non-trivial formal systems that would, indeed, satisfy this criterion are remarkably few in number. The most important of these are the propositional and predicate calculi of classical logic, which, as Badiou illustrates in the Appendix to *The Concept of Model*, *can* be demonstrated to be complete and consistent. The demonstration that Badiou exposits there is essentially that of Gödel's *completeness theorem*, but even this is far from providing an *internal* demonstration of the propositional calculus' completeness, calling as it does upon the arithmetical method of mathematical induction. The earliest non-inductive, and so effectively internal, demonstration of the completeness of the propositional calculus that I am aware of is due to the Dalhousie logicians Gilman Payette and Peter Schotch, and was arrived at only in 2004. For details, see Schotch, *Introduction to Logic and its Philosophy*, electronic resource, 2006, pp. 165-193. The text is available to download, free of cost, at <http://www.schotch.ca>.

26. Practically speaking, the requirement that *any* formally intelligible problem be soluble is a *heuristic impetus* guiding the creation of new axioms and not a criterion that measures axiomatic formalization's success (a criterion of which *almost* every attempt will fall short). It is nevertheless true that, in practice, a working formal system is capable of determining solutions for the better part of the problems that it can pose.

27. Immanuel Kant, *Critique of Pure Reason*, trans. Norman Kemp Smith, New York, Palgrave Macmillan, 2003, A727/B755.

> Archimedian geometry is something different in each case.[28]

Any notional residue that is not expressed in the axioms and which we might attach to the term 'point', for example, is not (or not yet) mathematical. Definitions, supernumerary with regard to a complete axiomatic system, have at best a psychological, heuristic value, and at worse re-entrench mathematics in its ideological base to the detriment of the former. As Hilbert remarks,

> If one is looking for other definitions of a 'point', e.g., through paraphrase in terms of extensionless, etc., then I must indeed oppose such attempts in the most decisive way; one is looking for something one can never find because there is nothing there; and everything gets lost and becomes vague and tangled and degenerates into a game of hide-and-seek.[29]

For these reasons, where we would expect to find definitions in *The Foundations of Geometry*, Hilbert confronts us with a mere classification of letters.[30] The text begins with the words:

---

28. David Hilbert, 'Letter to Frege dated December 29th. 1899', trans. Hans Kaal, in Gottlob Frege, *Philosophical and Mathematical Correspondence*, Gottfried Gabriel (ed.), Chicago, University of Chicago Press, 1980, 40. The thesis presented here is a particularly interesting one, and its full elaboration and defence is outside the scope of the current essay, though I intend to provide a thorough demonstration of it elsewhere. For now, some additional light can be shed on the matter by once more situating ourselves with respect to Kant, and saying that a mathematical concept is structured more like a Kantian *schema* than what he calls a 'concept' in the strict sense. Even Kant implicitly concedes as much at various points in the first *Critique*, though he does not make the schematic structure of mathematical concepts explicit.

29. Ibid, p. 39. It is enlightening in this context to contrast Hilbert's assertion that only in the full unfolding of the axiomatic is any concept 'defined' with Euclid's definition of 'point' as 'that which has no part;' a definition of which E. W. Strong rightly remarked that 'If there were no geometry in respect to which a point is defined, we should not know whether a point signified the holiness of an angel or the extremity of a line'. See Euclid, *Elements*, trans. Sir Thomas Heath, in Robert Maynard Hutchins, ed. *Great Books of the Western World*, Volume II, London, William Benton, 1952, Book One, Definition 1; and E. W. Strong, *Procedures and Metaphysics*, Hildesheim, 1966, p. 241. Quoted in A.T. Winterbourne, *The Ideal and the Real*, London: Kluwer, 1988, p. 69.

30. Even Hilbert's axiomatic geometry, however, ought not to be seen as the product of a 'complete formalization', an ideal towards which it clearly points. In *Leçons de logique algébrique*, Haskell Curry depicts this ideal as the reduction of the formal system to a level of strict, literal univocity. With this ideal in view, Curry writes that, 'for the strictest formalizations one seeks to reduce the morphology to a minimum'. Practically speaking, this means that rather than beginning with various, predetermined 'kinds' of elementary objects (points, lines and planes, for example), denoted by different species of marks (capital Roman letters, lowercase Roman letters, lowercase Greek letters), we begin with a single collection of uninterpreted letters and proceed from there. In short, what prevents Hilbert's geometry from acceding to the ideal that would only later be sighted by the formalists of his school, is precisely the classificatory schema to which definition has been reduced in Hilbert's text. What this ideal proposes is that the place of

> Let us consider three distinct systems of things. The things comprising the first system, we will call *points* and designate them by the letters, *A*, *B*, *C*, ...; those of the second, we will call *straight lines* and designate them by the letters *a*, *b*, *c*, ...; and those of the third system, we will call *planes* and designate them by the Greek letters α, β, γ, ...[31]

Note that it is not the use of 'symbols' or 'letters' as opposed to words that is of the essence here. A formalization, which fixes its words in a rigid apparatus well before any decision on notation is necessary, is a process entirely distinct from mere abbreviation—a distinction that is useful in discerning genuine formalisms from the mathemes that mimic them in appearance alone while preserving the entire rhetorical mobility of natural language.

Implicit in mathematical formalism is the thesis that the material inscription of mathematical thought—its reduction to bare letters and axiomatic rules of manipulation—is *a priori* adequate to its essence;[32] it is the thesis that thought can be made fully immanent to its mode of expression, such that it becomes pointless to distinguish between the accidents of its expression and the ideas expressed. In a single stroke, formalization attempts the materialization of thought and the idealization of its material support. Only by grasping this double movement can one understand how formalism allows, for example, the unmediated entrance of reasonings over the literal length of this or that formula into derivations that seem to bear on the mathematical structures that those formulae express (§ 8 of Badiou's Appendix to this book contains a derivation such as this). It is the ideal of an absolute adequacy of the spirit to the letter of mathematics that legitimates Hilbert's famous statement that

> in my theory contentual inference is replaced by manipulation of signs according to rules; in this way the axiomatic method attains that reliability and perfection that it can and

definition can be reduced further still. A classic example of such a system is the Zermelo-Frankel axiomatic of set theory, which begins with nothing but a single collection of variables and a binary predicate '$\in$' (which, if one chooses, may be read: 'belongs to'). The rest of the system, which is vast and complex enough to encompass *all* of classical mathematics (*i.e.*, all mathematics produced prior to the invention of set theory), is univocally deployed by the axioms.

31. Hilbert, *The Foundations of Geometry*, p. 3.

32. See David Hilbert, 'The Foundations of Mathematics', in Jean Van Heijenhoort (ed.), *From Frege to Gödel: A Source Book in Mathematical Logic, 1879-1931*, Cambridge, Harvard University Press, 1967, p. 464: 'I should like to eliminate once and for all the questions regarding the foundations of mathematics [...] by turning every mathematical proposition into a formula that can be concretely exhibited and strictly derived'.

> must reach if it is to become the basic instrument of all theoretical research.[33]

It is this basic tenet, more than anything else, that sets Hilbert at an irrecoverable distance from intuitionists such as L. E. J. Brouwer, for whom mathematical thought transcends the vagaries of its material support—or, to put it conversely (but equivalently): for whom the immanence of mathematical truth resists any sort of literal transcendence.[34]

Hilbert's understanding of what constitutes a scientific concept is paradigmatic for Badiou, who takes the very task of science to be the transformation of '*undefined* notions into concepts by way of their "axiomatic" liaisons within the system' (RMD 454). This Hilbertian thesis is mobilized against even Althusser himself (whose formal conception of science was peered by few), and used to cast doubt on the latter's 'identification of concept and definition' (RMD 464, n.28).[35] This identification, Badiou argues, throws us back into the fetters of an ideological representation of science that can no longer be upheld in the light of the formalist revolution in mathematics. We must recall, Badiou argues,

> that the concepts of a science are necessarily of undefined words; that a definition is never anything more than the introduction of an abbreviating symbol; that, consequently, the regularity of the concept's efficacy depends on the transparency of the code in which it figures, which is to say, on its virtual mathematization... (RMD 464, n.28)

The indifference that a formalized axiomatic bears toward its representational birthplace should not be mistaken for an *incompatibility* between the formal and the representational. 'Point', 'line' and 'plane', for Hilbert, do not cease to resonate with the geometrical imagination, on which he himself has written a fine book (whose title is simply: *Geometry and the Imagination*).[36] Nor does formalization separate mathematical domains

33. Hilbert, 'The Foundations of Mathematics', p. 467.

34. I offer an extended analysis of Brouwer's intuitionism and the relations it bears with Badiou's own theory of the subject in 'The Law of the Subject: Alain Badiou, Luitzen Brouwer and the Kripkean Analyses of Forcing and the Heyting Calculus', in Paul Ashton, A. J. Bartlett, & Justin Clemens (eds.), *The Praxis of Alain Badiou*, Melbourne, re.press, 2006, pp. 23-70. Incidentally, my analysis aims to demonstrate a *model-theoretic* confluence of the Badiousian and Brouwerian theories of the subject.

35. As evidence of this identification, Badiou quotes Althusser speaking of '... the question of the status of the *definition*, that is, of the concept', *Lire le Capital*, Paris, Presses Universitaires de France, 1965, vol. II, p. 67.

36. David Hilbert and S. Cohn-Vossen, *Geometry and the Imagination*, trans. P. Nemenyi, New York, Chelsea Pub. Co., 1952.

from all semantic considerations—on the contrary, it is absolutely prerequisite to the rigorous treatment of model-theoretic questions, whose solutions seldom fail to shed light on the formal systems in question. The point is that formalization allows mathematical practice to achieve an *indifference* to representation, and it is for this reason that Badiou takes formalization to be the essence of mathematics' break with ideology. This identification allows Badiou to posit the existence of a radical break between the scientific and the ideological while simultaneously affirming that science has no other terrain, and no other raw material, than ideology (whether ideology in general or the autochthonous ideology given in science's self-representation). It is easy to see how, with Badiou's early epistemological studies into mathematical formalization, a paradigm is set which remains dominant through the entirety of his work. It is here that the crucial notion of 'subtraction' is to be found.

We cannot stop here, however, if we are to arrive at anything other than sheer epistemic asceticism. Subtraction alone gives us very little; it certainly does not give us anything new. To complete the category of the epistemological break, to lay hold of its positive content, it is necessary to return to the figure of 'remaking',[37] the dimension of 'retroactive causality' that is characteristic of the epistemological break insofar as the latter is seized as the 'history of formalization'. Badiou's clearest early treatment of this aspect, outside of *The Concept of Model*, is a text published in the ninth issue of *Cahiers pour l'analyse*, in the same year as Althusser's Course, bearing the title, 'La subversion infinitesimale' ('Infinitesimal Subversion').[38]

The text has both a general and a particular aim. Its first part is concerned with introducing a new category into the structuralist epistemology of mathematics: that of 'infinity-points'; the second takes up a detailed case study of the concept of infinitesimal quantity, from its awkward entrance into the world of mathematics in the infinitesimal calculus, to Bishop Berkeley's decisive critique in *The Analyst*, and Cauchy's utter elimination of the concept in favour of a set-theoretical theory of limits, to its then recent resurrection in Abraham Robinson's rigorous

37. Following Robin Mackay's working translation of 'La subversion infinitesimale', on which I have heavily relied in preparing this text, I have translated '*refonte*' by 'remaking'. See Alain Badiou, 'Infinitesimal Subversion', trans. Robin Mackay, in Peter Hallward and Christian Kerslake (eds.), *Concept and Form: The Cahiers pour l'analyse and Contemporary French Thought*, forthcoming, 2010; details available at http://www.mdx.ac.uk/www/CRMEP/varia/Cahiers.htm.

38. Alain Badiou, 'La subversion infinitesimale', in *Cahiers pour l'analyse*, vol. 9, 1968. Henceforth cited SI.

formalization of non-standard analysis.

Badiou initially situates the category of 'infinity-points' in the context of the Turing Machine—more for the purpose of giving the idea greater tangible clarity, and for underscoring the materiality of place and mark essential to Badiou's conception of formal thought, than for any reason specific to Turing's formalism. To summarize, we may, for the sake of discussion, assume the existence of a given consistent formal system, call it X, endowed with a set of operations; we then imagine this system as expressed in Turing Machine, such that every operation results in the marking or erasure of a certain, specified place on an infinitely long tape. For each operation *o* we assume a place on the tape that cannot be marked (or erased) as a result of any application of *o* over its admissible values. An *infinity-point* for some operation is then defined as a supplementary mark determined in such a way that it '(a) occupies the unoccupiable empty place; (b) apart from this occupation, follows all of the initial procedures' (SI 119). In the case of the operation of *succession*, this is fairly simple to grasp: its infinity point would be some mark, we could call it I, such that no *n* exists that has I as its successor, and such that it is possible to define the successor of I. We should not read too much into the term 'infinity-point', however: an equally valid example would be the 'imaginary number' *i* , defined as the impossible square-root of –1. (The reader may note that these are precisely the terms commonly referred to as 'ideal elements'; the change in terms reflects an ideological shift, which is by no means negligible.)

The creation of a new infinity point in a given domain is no trivial matter; it takes place on the very cusp of the domain's rationality, which it threatens to tip into a maelstrom of contradiction, and once accomplished effects a radical *remaking* of the entire domain, which is always, Badiou insists, 'a theoretical violence, a subversion' (SI 128). Let me approach these dangers one at a time. The first—the threat to rationality as such involved in the inscription of an infinity point—bears on the very concept of logical consistency. The concept of logical consistency that Badiou adopts is informed by the definition formulated by the great American logician, Emil Post. In contrast to the traditional definition of consistency as non-contradiction (a system is consistent if is does not allow the derivation of a statement *and* its negation), Post identifies a system's consistency with the existence of a statement that cannot be derived for that system, a statement whose derivation is impossible (notice that, where the logic is classical, Post's formulation implies the traditional one, and is in fact equivalent to it, but this need not be the case elsewhere). In the context of Turing's formalism, this means that a system is consistent if and only if there exists a place that cannot be marked (or erased). In a surprising but enlightening simplification of Lacanian doctrine, Badiou then poses the following:

> for a domain of fixed proofs, the real is defined as the impossible. It is owing to the exclusion of certain statements, the impossibility of having the constants occupy certain constructible places, that an axiomatic system can operate as the system that it is, and allow itself to be thought differentially as the discourse of a real. (SI 122)

The alternative to such determination by the real is for the system to be inconsistent, in which case 'every statement is derivable [...] all constructible places are occupiable and the system no longer marks either differences or regions, makes itself an opaque body, a deregulated grammar, a language thick with nothing' (SI 122). There is thus always the fear that the inscription of an infinity-point will throw the system into the brown fog of inconsistency, for,

> since it is linked to the forcing of the empty spaces proper to a domain, the introduction of an infinity-point is a modification [*remaniement*] which must seem irrational, since in any given theoretical conjuncture rationality is defined precisely in respect of these blank spaces, the sole assurances, variably indexed, of a *real* difference in the domain.

The beauty of formalizing the Lacanian real in this way lies in its sheer simplicity, and the force with which it strips away the cobwebs of mystification and esotericism. It is an oversimplification of Lacanian doctrine, to be sure, but one that constitutes the primitive logical skeleton—the 'rational kernel'—of the notion of the real that endures through Badiou's writings, whatever obscure patina it acquires with time.[39] It permits us, in effect, to make sense of the cryptic mantra that runs throughout 1982's *Théorie du sujet*, where Badiou responds to Lacan's dictum that 'the real is the impasse of formalization', with his own maxim that 'formalization is the place where the real passes in force' (TS 40), a mantra by which Badiou encapsulates all the difference between a 'structural', 'idealist'

39. Badiou ratchets the logical complexity of the category of the real up a notch in *Being and Event*, where he identifies the real not only with the indemonstrable, but with the *undecidable*—that subclass of the indemonstrable that can neither be proved *nor disproved* within a given axiomatic. See § 2 of Badiou's Introduction to *Being and Event* for a brief discussion of this shift.

dialectic that treats the real only as an impasse, a fissure, an impossibility that cannot be broached, and a 'historical', 'materialist' dialectic that treats the real as the forced passage of a new formalization at the point where an impasse is reached.

The action of 'remaking', contingent on the occupation of the real through the inscription of an infinity-point, is precisely such a forcing of the real. Involved in the act of 'remaking' is the supplementation of a series of axioms to the existing system correlated to the introduction of its infinity point, a supplementation that is by no means guaranteed success (it is, after all, a trespass against the real). These axioms not only determine the behaviour of the newly introduced 'infinity-points' but proceed to effect a general transformation of the structure of the domain. For instance, when Abraham Robinson introduces, into the field of real numbers $\mathbb{R}$, an element α defined in such a way that $n < \alpha$ for every $n \in \mathbb{R}$, and supplements the axiomatic of this field with a series of new axioms regulating the operational rules pertaining to α (rules involving, above all else, the definition of infinitesimals as the result of dividing reals by α), the resulting field is transformed, not only in the particular elements it allows, but in its very structure: the Archimedian field of reals is remade into a non-Archimedian field of reals and infinitesimals (so that the rule $(\forall m,n)(\exists k)(km > n)$ no longer holds for the remade domain).[40]

The action of 'remaking', in sum, is the transformation of an existing domain through the supplementation of a new formalization that breaches the limit of what was previously impossible, or unthinkable in that domain—a breach symbolized (literally) by the inscription of an infinity-point in the field of the real. Readers familiar with Badiou will hear, in this, a dozen echoes of his later texts. We will take up one such recurrence in due course, when we come to the concept of forcing as it stands in *Being and Event*.

III. On Objectless Materialisms

Badiou enlists mathematical formalization's capacity to unshackle itself from the representational domains from which it proceeds in a series of novel ruminations on mathematics' *non-objectal* character. What is made

40. Complete technical details are provided in Badiou, 'Infinitesimal Subversion'. Also see my translation of 'Mark and Lack', trans. Zachary Luke Fraser, in Peter Hallward and Christian Kerslake (eds.), *Concept and Form: The Cahiers pour l'analyse and Contemporary French Thought*, forthcoming, 2010.

evident by formalization is that, in mathematics, 'nothing authorizes the determination of an object. Here, the thing is null: no inscription can objectify it' (MM 156). Even the representational points of departure offer no obstacle to this conception, for no sooner do they enter into relation with the products of formalization—no sooner do they reconfigure themselves as *models*—than they are themselves transfigured into sheerly scriptural apparatuses. This touches on one of the major themes of the present book, which I will no more than indicate in this essay: what a careful study of the formalizations presupposed by any consideration of 'models' reveals is the incorporation of *both* formal syntaxes and semantic structures into a single mathematical situation, a situation exhausted in the deployment of notational differences.

The untethering of mathematics from the figure of the object may well seem to send it drifting irretrievably into the fog of idealism. It is interesting that the conclusion Badiou draws from this process is just the opposite: in breaking with the category of object, mathematics becomes the 'anti-idealist exercise par excellence', for 'the knot of idealism is not the category of the subject but that of the object'.[41] The subject, for its part, can only be adequately made the topic of a rigorously materialist philosophy in the wake of the object's dissolution—hence Badiou's remark, in the seminar, conducted at Paris University VIII in 1987, that 'a materialist (and therefore objectless) theory of the subject is necessarily of the school of post-Cantorian mathematics'. We should be cautious here, however, and avoid being carried along too quickly. That a constant thread of 'non-objectal mathematical materialism' runs from one end of Badiou's work to the other is no excuse for ignoring the profound transformations of context that take place in the course of its development. Two crucial changes must be taken into account, before we move any further: one is the appearance of the subject in Badiou's later materialism, whose disappearance in the Althusserian epistemology that Badiou develops in the '60s we have barely registered as yet (we will come back to this soon); the other is the shift from a preoccupation with formal mathematics and mathematical logic—with what we could call the school of post-Hilbertian mathematics—to mathematical set theory, 'the school of post-Cantorian mathematics'. Between these two movements there is both contentual and historical overlap of considerable degree—there is

41. Alain Badiou, 'Orientation de pensée transcendante', a seminar conducted on 24 October 1987 at Paris University VIII. I thank François Nicolas for providing me with the notes that he had compiled from this seminar.

no doubt about this—but their bearing in Badiou's philosophy changes dramatically. Badiou illustrates this point himself in his preface to Fayard's recent republication of *Le concept de modèle*, where he remarks that, between *Le concept de modèle* and *L'être et l'événement*,

> I no doubt pass from a positive reading of mathematics as the place of ruled inscriptions to a reading according to which the mathematics of the multiple is the thinking of being qua being. In brief: I pass from a structural materialism, which privileges the letter (the mark), to an ontological materialism, which privileges the evidence of the 'there is', in the form of pure multiplicity, mathematically reflected for the first time by Cantor. Certainly, this evidence, like every mathematized evidence, is bound to the letter, but it cannot be reduced to it.[42]

We will tackle these variations in turn, while paying close attention to the interweaving they demand between materialism and formalization.

IV. Structural Materialism

The materiality in question in Badiou's 'structural materialism' is that of the *letter*, which grounds both the materiality of mathematics and its independence from mental particularities (its 'objectivity').[43] But what is the letter, and in what sense is it *material*?[44] Here, we're faced with some degree of difficulty, as the simplest explanation is immediately disqualified: The materiality of the mathematical letter *cannot* consist in its physical instantiation, in the tangible deposits of graphite or chalk; this much must be clarified at once. Mathematics, even as seen through a formalist lens, is not the empirical science of specific scriptural marks, despite certain formulations of Hilbert and his school.[45] Faced with an assemblage

42. Alain Badiou, 'Préface de la nouvelle édition', *Le Concept de modèle*, Paris, Fayard, 2007, p. 27-28.

43. For a brief but illuminating account of the function of 'the letter' in Badiou's later work, see Justin Clemens, 'Letters as the Condition of Conditions for Alain Badiou', in *Communication & Cognition*, vol. 36, no. 1-2, 2003, pp. 73-102.

44. Let us note straightaway that words, too, taken as units, can serve as 'letters', as can any other conceivable mark or sign. The term 'letter' is being used here in a rather synechdocal sense. It is not uncommon for 'word' to be used similarly. Such use is common in the writings of logicians, where one often speaks of 'logical words' and 'sentences' to refer to connectives and formulae.

45. For example, in an important lecture given in 1927 and entitled 'The Foundations of Mathematics', Hilbert explained that 'as a condition for the use of logical inferences and the

of tangible inscriptions, all that matters to the mathematician, qua mathematician, are the (intangible) rules for manipulating these marks and the (incorporeal) network of differences that inhibits the confusion of those rules. So long as these non-physical structures remain intact, their physical support may undergo any sort of modification without appreciable effect.[46] The system deployed in *The Concept of Model*, for example, would be entirely unaffected by a substitution of the signs ∀ and ∃ for the U and E used in the original text, or the italicization of the letters standing for variables and constants—alterations which I have, in fact, made in the course of my translation. However, a substitution that did not preserve differences—if, for instance, I used ∃ for *both* U and E—would of course be disastrous.[47]

This insight, which is central to the doctrine of 'scriptural materiality'—a doctrine that, Ray Brassier argues, constitutes much of 'the esoteric subtext of Badiou's materialist epistemology of science'[48]—owes much to Ferdinand de Saussure's *Course in General Linguistics*, a text that had a profound influence on the 'structuralist' academia of 1960s Paris.[49] In it,

performance of logical operations, something must already be given to us in our faculty of representation, certain extralogical concrete objects that are intuitively present as immediate experience prior to all thought', Hilbert, 'The Foundations of Mathematics', in *From Frege to Gödel*, p. 464.

46. This illustrates one of Saussure's central theses, which is that the *differential* character of the sign is intimately related to its *arbitrary* nature: '*Arbitrary* and *differential* are two correlative qualities'. See Ferdinand De Saussure, *Course in General Linguistics*, trans. Wade Baskin, Toronto, McGraw-Hill, 1959, p. 118

47. A second 'decorporealizing' determination affects the mathematical sign by way of its very rules of production, as Mary Tiles has brought to light in *Mathematics and the Image of Reason*, in a passage which, although it falls to the margins of the current essay, deserves to be quoted at length: 'The idea of a pure formal calculus, an uninterpreted notation, is that of a system generated by a set of rules for producing sequences of marks on paper, where it is possible to specify an algorithm which will determine whether any given sequence has or has not been produced in accordance with the rules. [...] Because rules are repeatedly applicable they are already inherently general. Rules which are rules of production, construction, determine the character of the product (are constitutive) in just those ways which make it possible to tell from the product whether it was or was not constituted according to the rules. The kind of rules thought to characterize a formal system thus immediately traverse the gap between particular and universal, token and type. In this way concrete marks cannot remain without signification; they symbolically signify the types of which they are the tokens. [...] To ask whether '0 = 1' is provable within a particular formal system of arithmetic cannot be to ask whether the particular token inscription on the previous line can be so proved – it clearly was not; it is to ask whether the formula (type) is provable', p. 131-2. Consequently, Tiles reasons, 'Numerals are no less abstract than numbers', p. 132.

48. Ray Brassier, 'On Badiou's Materialist Epistemology of Mathematics', p. 145.

49. The sort of 'structuralism' emanating from Saussure's work is not *directly* connected to the mathematical 'structuralism' epitomized by the Bourbaki group, though some cross-fertilization did occur. Michel Serres' philosophical work is a fascinating instance of this sort of thing. Ba-

Saussure develops his fundamental insight that the signifying web of language has its fundamental support not in the physical qualities of words themselves, but in 'the phonic differences that make it possible to distinguish this word from all the others, for differences carry signification'.[50] The linguistic signifier must therefore be understood as 'not phonic but incorporeal—constituted not by its material substance but by the differences that separate its sound-image from all others'.[51] And so, Saussure concludes, 'in language there are only differences *without positive terms*'.[52] This view of things exerted enormous influence over the Badiou of 1968, as can be seen throughout the current book. It can be seen, for example, in his assertion that 'mathematical experimentation has no material place other than where difference between marks is manifested' (CM 30), a thesis he arrives at upon observing that all that need distinguish a formal system from the structure in which it is interpreted is a mere notational difference—in the current context, the difference between the constants of the language, *a, b, c*, ... , and the 'objects' of the universe, *u, v, w*, ...

The formalizing rigours of Hilbertian mathematics make reference to any 'external object' superfluous, and return us instead to the immanent materiality of mathematical inscriptions, but, far from offering up the solid bedrock of an objective point of reference, it seems that the literal 'materiality' at stake is at bottom an incorporeal web of sheer differences, instantiated in chalk and graphite but determined elsewhere entirely. At this point, any effort to call 'materialist' a project that stakes its foundations in an objectless plane of incorporeal differences is bound to appear rather unorthodox, to say the least. I will leave aside, for now, the hoard of questions that the very notion of 'incorporeal materiality' conjures up. They cannot receive the attention they deserve in this essay.

V. Ontological Materialism

Badiou's thesis, first formulated in 1984's 'Custos, quid noctis?' (a short review of Lyotard's *Le Différend*) and elaborated at length in 1988's *L'être et l'événement*, that mathematics itself is the science of being qua being, constitutes one of the most difficult and problematic points of his mature

---

diou's early work, to some extent, is another.

50. Saussure, *Course in General Linguistics*, p. 118.

51. Saussure, *Course in General Linguistics*, pp. 118-9.

52. Saussure, *Course in General Linguistics*, p. 120.

philosophy. Needless to say, the discussion of it that I offer here will be somewhat abbreviated. It will be enough to demonstrate, however, that the problems raised by the 'binding' of being to the letter are problems that the present book makes difficult to ignore.

The simplest angle from which to approach Badiou's identification of mathematics and ontology is with respect to the restricted form that predominates in *Being and Event*: the narrower identification of set theory as the science of being, insofar as it constitutes a situation in which the disclosure of the structure of the 'there is' takes place through the axiomatic deployment of pure multiplicity. With reckless abridgement, we can summarize the notion of the 'there is' at play in this thesis by referring to its instantiations in a given 'states of affairs', in what Badiou calls 'situations', or, more abstractly, 'presentations'. While we may speak, in each case, of *a* presentation, or *a* situation, intrinsic to Badiou's notion of the 'there is' is that the unity by which we grasp it is extrinsic to its sheer occurrence: 'there is no one, only the count-as-one. The one, being an operation, is never a presentation', we are told at the outset of his book (BE 24). The non-inherence of unity in presentation leads Badiou to characterize the latter as *multiplicity*, which, in turn, receives its rigorous conceptualization in the mathematical figure of the *set*.[53]

It is often said that the non-intrinsic-unity of multiple-presentation is what prevents us from calling it an 'object', but there is no reason why this should be the case. That objects are only unitary in virtue of an operation of unification extrinsic to their being is a fundamental tenet of Kantian philosophy, whose conception of the object is not at all foreign to Badiou. In fact, when Badiou does come around to developing a positive concept of 'object' (which is lacking from all his previous texts[54]) in 2006's *Logique des mondes*, he defines it in a strikingly Kantian manner,

53. There exists a notable anticipation of Badiou's later identification of set theory and ontology in that mother structure of his future oeuvre, 'Le (Re)commencement du matérialism dialectique', where we read:

> There must exist a *preliminary* formal discipline, which we may be tempted to call *the theory of historical sets*, and which would *minimally* consist in the protocols of 'donation' of pure multiplicities over which structures would be progressively constituted. This discipline, strictly dependent through its entire development on the mathematics of sets, would no doubt go beyond the simple donation of a procedure of *belonging*, or an inaugural system of empty differences. (RMD 461)

Note that 'historical sets' translates the French 'théorie des ensembles historiques', an allusion to both the French term for 'set theory' ('*théorie des ensembles*') and to the subtitle of Sartre's *Critique of Dialectical Reason*: '*théorie des ensembles pratiques*' ('theory of practical ensembles').

54. Naturally, this absence has proved to be a significant obstacle in trying to say anything intelligible on the topic.

as a multiple counted-for-one and submitted to the transcendental laws of appearance in a determinate world (the only thing un-Kantian about this is the absence of a 'transcendental subject' in which those laws would be inscribed). It is probably better to link the non-objectality of the multiple (which is invariably counted-as-one, even in mathematics, without for all that becoming an 'object') to its *ontological* rather than *ontic* status. This seems to be suggested by Badiou's remark in the 1987 seminar that 'the multiple in itself is not an object; it's the general form of the exposition of being', if we interpret the second clause as *implying* the first. The same seems to be said in the following passage, taken from the introduction to *Being and Event*:

> If the argument I present here holds up, the truth is that *there are no* mathematical objects. Strictly speaking, mathematics *presents nothing*, without constituting for all that an empty game, because not having anything to present, besides presentation itself—which is to say the Multiple—and thereby never adopting the form of the ob-ject, such is certainly a condition of all discourse on being *qua* being. (BE 7)

In any case, the next point that must be taken up is that, presentation being irreducibly multiple, there is no single, all-encompassing presentation or situation; there is no *totality* of 'what there is'. Ontology cannot subsume presentation, or even refer to being-in-totality as its object, and can therefore exist only as a particular situation. This situation is precisely mathematics as it historically presents itself—particularly, for the purposes of *Being and Event*, as it is concentrated in Zermelo-Fraenkel set theory.

Problems appear as soon as we try to make this notion more precise. A plausible interpretation of Badiou's characterization of set theory as the presentation of presentation would be that we should understand set theory as the situation in which the formal structure of any given presentation is presented, and that these formal structures are, simply, sets. This raises the question, however, of how one is to establish a correspondence between the 'sets' presented in the ontological situation, and other, 'concrete' presentations presented elsewhere.

This was, in essence, the question that arose from my own initial reading of Badiou's text. It seemed to me that there could be no hope of set-theoretic ontology providing the groundwork of a theory of situations and subjective actions without at least the possibility of a bridge between the two being elucidated. The question itself was intertwined with another one, which essentially was that if it is not a question of drawing a 'correspondence' between presentations and sets, if presentations simply *are* sets, then what are we to make of the statement in *Being and Event* that reads:

> The thesis that I support does not in any way declare that being is mathematical, which is to say composed of mathematical objectivities. It is not a thesis about the world but about discourse. It affirms that mathematics, throughout the entirety of its historical becoming, pronounces what is expressible of being qua being. (BE, 8/14)

This statement seems to directly contradict any supposition that Badiou claims sets and presentations to be precisely the same thing. But if their correlation is not one of identity, what is it? When I posed this question to Badiou in a letter written in 2005, I received the following response:

> The difficult point in your question is the sense of the word 'correspondence'. In my theoretical apparatus [*dispositif*], I believe that the question, 'what is the correspondence between a being and its set' is deprived of signification. In effect, only the notion of set gives sense to that of being, in the context of an ontology—so that it must be posited that 'all being' [« *tout l'être* »] is thought in and by the set. This does not at all mean that a given multiplicity is 'the same thing' as the set, but only that insofar as one thinks it in its being, one thinks it as a set.

Everything hinges here on a slight displacement: whereas I (conflating the ontological and the foreign dimensions singled out above) posited sets as being a certain kind of presented thing—so that we have one situation populated by sets and others populated by dogs, thieves, stars and syllables—Badiou implies that the set is *not* one kind of thing among others, but neither is it the substance of all things. It is precisely a mode of thought; the aspect that is produced of each thing when that thing is thought in its being is exactly a *set*. No situation, no presentation is composed of sets; rather, set theory deploys the means by which any presentation may be thought *as* a set, and this mode of thinking is by definition 'ontological'.

In order to see what this could mean concretely, we must ask what exactly the ontological situation might be—if it is not, strictly speaking, a presentation of sets. Abstractly speaking, it is a presentation of presentations, like everything else. This of course tells us nothing; and the real question concerns how these presentations are structured, or counted-as-one. Roughly speaking: what do we find there, and how would we recognize it? The answer now seems simple enough: Faithful to Hilbert, we will say that it is not at all a presentation of 'sets', but of *signs*. Paraphrasing Quine, we will then say: *for something to be thought in its being for it is to be thought as the value of an ontological variable.*[55] To think a particular structure 'in its being'—that is, according to its pure presentational form—is simply to think it in the symbolic language of set theory, which is one and the same as thinking it 'in and by the set'. Hence, it is not the case that 'sets' are brought into correspondence with situations or 'structures', but that structures are brought into correspondence with *symbolic expressions*, both terms belonging to the same general category (the category of the existent; that is, of the presented, the consistent, the counted-as-one). The concept of 'set' itself is, qua concept, only quasi-mathematical—as Badiou notes, nowhere in set theory itself does the concept of 'set' play any operative role. It arises only as a sort of phantasm whenever the axioms of set theory are satisfied in some domain (a phantasm that is no doubt both useful and misleading in guiding mathematical intuition).[56] For example, since the demonstration of the Löwnheim-Skolem theorem, we have known that it is possible to interpret the axioms of ZF in such a way that they are satisfied in a model constructed solely out of natural numbers. In such circumstances, this numerical domain is truly thought in and by the set.

If this is how we are to understand things, what this means is that the relation between ontology and the various concrete situations that it 'thinks' is to be conceived as the relation between a syntactic apparatus and the models that satisfy it. This seems to be how Badiou's use of mathematical ontology plays itself out in *Being and Event*: in his employment

55. Quine's original formulation—'to be is to be the value of a variable'—is, in fact, cited by Badiou himself, in the context of discussing set-theory's ontological univocity:

> If we admit—with a grain of salt—Quine's famous formula: 'to be is to be the value of a variable', we may conclude that the system ZF postulates that there is only one type of presentation of being: the multiple. (BE 44)

The original source of Quine's aphorism is his essay, 'On What There Is', in *From a Logical Point of View: Nine Logico-Philosophical Essays*, Cambridge, Harvard University Press, 2006.

56. Jean-Louis Krivine put it wonderfully in his, *Théorie axiomatique des ensembles*, Paris, Presses Universitaires de France, 1969, where he describes set theory as 'the theory of binary relations satisfying the Zermelo-Fraenkel axioms' p. 6. Accordingly, a set is nothing but a term of a relation satisfying these axioms. Now, it may fairly be said that this just shifts the bump in the rug, since we have only replaced the notion of 'set' with that of 'relation' (a move which Badiou is rather hostile to, incidentally).

of both Gödel's notion of constructibility and Cohen's notions of genericity and forcing, Badiou explicitly treats the situations in question as *models*.

It is here that Badiou's first book comes home to roost, with its troubling demonstration that, in any case, it is only possible to treat a domain as a model for a mathematical syntax if that domain is *already mathematical*—if, for instance, it is already organized along set theoretic lines. In *The Concept of Model*, Badiou shows how

> the concept of model is strictly dependent, in all its successive stages, on the (mathematical) theory of sets. From this point of view, it is already inexact to say that the concept connects formal thought to its outside. In truth, the marks 'outside the system' can only deploy a domain of interpretation for those of the system within a *mathematical envelopment*, which preordains the former to the latter. [...] Semantics here is an *intramathematical* relation between certain refined experimental apparatuses (formal systems) and certain 'cruder' mathematical products, which is to say, products accepted, taken to be demonstrated, without having been submitted to all the exigencies of inscription ruled by the verifying constraints of the apparatus (CM 42).

And so the ontological/mathematical formalization of situations is possible only in the light of their pre-ontological mathematization. This is something that Badiou hints at from time to time, but which he has never philosophically thematized. This notion seems implicit, for instance, when he speaks of a 'horizon of mathematicity' in which *any* situation in disclosed, and upon which mathematical physics, for example, operates.[57]

The difficulties of the situation in which Badiou places himself in *Being and Event* are considerable, but they have not been fruitless. As we see in the interview included in this volume, it was precisely the sort of impasses caused by the unintelligibility of the relation between the specifically ontological situation and situations in general that spurred the ambitious constructions undertaken in the 'sequel' to *Being and Event*, *The Logics of Worlds*. As we will return to these difficulties in the interview, there is no need to dwell on them any longer here.

---

57. Alain Badiou, *Ethics: An Essay on the Understanding of Evil*, trans. Peter Hallward, London, Verso, 2001, p. 128.

VI. 'The Chief Defect of All Hitherto Existing Materialism'

There is, in the end, a tendential, or perhaps only approximate, convergence between the structural and ontological varieties of materialism in the category of the letter. Little is gained by leaving things at that, however. We should go further.

The knot that ties together the two, disparate sheaves of objectless materialism that Badiou puts forth, can, perhaps, be found in Marx's own canonical proposal concerning the separation of materialist philosophy from the category of the object. The proposal in question is his first thesis on the philosophy of Feuerbach, where he argues that

> [t]he chief defect of all hitherto existing materialism—that of Feuerbach included—is that the thing, reality, sensuousness, is conceived only in the form of the object or of *contemplation*, but not as *human sensuous activity, practice*, not subjectively. Hence it happened that the *active* side, in contradistinction to materialism, was developed by idealism—but only abstractly, since, of course, idealism does not know real, sensuous activity as such.[58]

This leads us back quite naturally to the *act* of formalization itself, and to the practical processes that formalization determines. Badiou's most eloquent (but elliptic) pronouncement on the connection between formalization and the act takes place in the twelfth lesson of *The Century*, where, first, in a meditation on the artistic avant-garde, and again, in a reflection on mathematical formalism and the Bourbaki project, he remarks on the twentieth century's need to lay hold *of the act of thought itself*, beyond any consideration of content or representation. This is accomplished through *form*, and through form alone. Here, however,

> we need to contrast two senses of the word 'form'. The first, traditional (or Aristotelian) sense is on the side of the formation of a material, of the organic appearance of a work, of its manifestation as a totality. The second sense, which belongs

58. Karl Marx, 'Theses on Feuerbach', in Robert C. Tucker (ed.), *The Marx-Engels Reader*, New York, W. W. Norton, 1978, p. 143. The second clause of Marx's thesis, moreover, suggests why intuitionism cannot serve as a materialist doctrine in this context: while intuitionism heroically preserves the act while eschewing the category of object as an artefact of reification, it does so in a profoundly and purposefully idealist manner. This is true to such an extent that every relation that the Mathematical Act might entertain with matter, language, or other human beings is considered to be superfluous to its essence and harmful to its practice. At best, 'material existence' is a necessary evil against which the purest intuitionists must struggle in a manner comparable with the mystics (with whom Brouwer always had great sympathies).

> to the century, sees form as *what the artistic act authorizes by way of new thinking*. Form is therefore an Idea given in its material index, a singularity that can only be activated in the real grip of an act. Form is the *eidos*—this time in a Platonic sense—of an artistic act; it must be understood *from the side of formalization*. [...] But in 'formalization', the word 'form' is not opposed to 'matter' or 'content', but is instead coupled to the real of the act.[59]

What is true of art is true of mathematics as well, to which Badiou dedicates a number of passages worthy of repeating here. I will limit myself to the following: a passage in which he proclaims the singular importance of Hilbertian formalism, as expressed, in particular, by the Bourbaki group of French mathematicians. What Badiou singles out as worthy and demanding of our attention is the formalists' effort

> 'to break in two', as Nietzsche would say, the history of mathematics, in order to establish a comprehensive formalization, a general theory of the universes of pure thought. To produce in this manner the steadfast certainty that every correctly formulated problem can be solved. To reduce mathematics to its *act*: the univocal power of formalism, the naked force of the letter and its codes. Bourbaki's great treatise is France's contribution to this cyclopean intellectual project. It is necessary to lead everything back to a unified axiomatic; to compel formalism to demonstrate its own coherence; to produce—once and for all—the 'mathematical thing', never abandoning it to its piteous and contingent history. Everyone must be offered an anonymous and complete mathematical universality. The formalization of the mathematical act is the enunciation of the mathematical real and not an *a posteriori* form stuck onto an unfathomable material.[60]

It is in this sense that, in a strange but discernible Marxian fidelity, Badiou seizes upon formalization as the royal road to materialism—it produces 'the thing' as pure act, captured in 'the naked force of the letter', and dissolved in the insubstantial univocity through which mathematics renders itself the science of being.[61]

---

59. Alain Badiou, *The Century*, trans. Alberto Toscano, London, Polity, 2007, pp. 159-60.

60. Badiou, *The Century*, pp. 162-3.

61. 'Mathematics is a thought, a thought of being qua being. Its formal transparency is a direct consequence of the absolutely univocal character of being. Mathematical writing is the transcription or inscription of this univocity', Alain Badiou, 'Notes on Being and Appearance', in

We have yet to unfold the details of the formalizing act, as Badiou understands it. Throughout Badiou's work, the theoretical articulations of the formalizing act are manifold; we will single out two, which stand out by the force of their originality and the sophistication of their conceptual constructions. These same texts, significantly, take up the very problem that we have left in the dock a few pages back: the problem of how we are to conceive of science's subtraction from ideology. The endurance of these problems is all the more striking when measured against the utter dissimilarity of the two texts, separated in date by almost twenty years: 1969's 'Marque et manque' and 1988's *L'être et l'événement*. I will take them up one at a time, but not in isolation—a handful of other texts will have a part to play in shedding light on these writings.

VII. From Machinic Psychosis to Subjective Fidelity: 'Marque et manque', and *L'être et l'événement*

Written one year before the delivery of *The Concept of Model*, 'Mark and Lack' undertakes two essentially connected projects: first, to produce a radical critique of attempts by Jacques-Alain Miller and Jacques Lacan to make mathematics an object for psychoanalysis; second, to refine the category of the epistemological break in such a way as to give a precise sense to Lacan and Miller's errors. The focal point of the essay is the concept of *suture*, a concept which Miller found to be only implicit in Lacan's work and which he undertook to explicate in an essay of the same name. 'Suture', writes Miller, 'names the subject's relation to the chain of its discourse. One will see that it figures there as the lacking element, in the form of a placeholder'.[62] The central object of criticism in Badiou's text is Miller's effort in that text to exploit the concept of suture in an analysis of Gottlob Frege's *Foundations of Arithmetic* (an effort remarkable for its originality, whatever its flaws).[63] Through the course of his analysis, Miller tries to demonstrate that Frege, in order to found arithmetic in logic, necessarily has recourse to the lacking element that marks the suture of the subject, an element which, in Frege's text, falls under the

---

Theoretical Writings, ed. and trans. Ray Brassier & Alberto Toscano, London, Continuum, 2003, p.173.

62. Jacques-Alain Miller, 'Suture (elements de la logique du signifiant)', *Cahiers pour l'analyse*, no. 1, 1968, p. 41.

63. See Gottlob Frege, *Foundations of Arithmetic, A logico-mathematical inquiry into the concept of number,* trans. J.L. Austin, New York, Harper, 1960.

name of *the non-self-identical*, an element that is 'summoned and then annulled' in the Fregean definition of zero. (Zero being the number of the extension of the concept 'not identical with itself'.) Miller argues that this definition has the effect of surreptitiously grounding arithmetic in the subject (the 'non-self-identical'), which mathematics must invoke for the sake of its commencement and exclude for the sake of its consistency. If the non-identical subject were not symbolically annulled at the very instant of its invocation, Miller argues, there would be an utter collapse of the mathematical 'field of truth', in which identity and substitution *salva veritate* are one and the same. The mark '0', we are told, indicates the site of this double movement, or 'suture' of the subject.

Badiou's attack on Miller's text is thorough, and severs the very root of the Lacanian's argument: mathematics, Badiou argues, neither 'summons' nor 'annuls' anything which it does not itself produce, and that in this 'double movement' there is no paradox at all, since there is always a difference in level, or *strata*, between the production of a formula and its deductive refutation. In the example in question, this means that there is no mysterious contradiction or duplicity to be found between the production of the predicate 'not identical with itself' and the proof that this predicate's extension is void. Crucially, Badiou's essay illustrates how the stratified texture of mathematical discourse is evident only through its thorough formalization, which requires us to distinguish sharply, for example, between the mechanisms of concatenation ('mechanism-1'), formation ('mechanism-2'), and derivation ('mechanism-3') operative in a logical calculus. We cannot confuse, that is to say, the stringing together of the marks of the logic (under mechanism-1) with the sorting of these strings into well-formed and ill-formed (under mechanism-2, nor can we confuse the well-formed or ill-formed character of a formula with the derivation of its proof or disproof (under mechanism-3). Confusions of this nature always threaten to cloud logic with the greatest misunderstandings, leading one to posit, for example, a contradictory double-movement in which the 'non-identical' is 'summoned and annulled' in the demonstration that the extension of $(x \neq x)$ is empty. A clearer view of the matter reveals no double and paradoxical movement, but two *distinct* movements on different strata and under different mechanisms: the production of the well-formed formula $(x \neq x)$ by mechanism-2 and the derivation of its negation by mechanism-3: 'No absence is convoked here that is not the distribution, into one class rather than its complement, of the productions that this mechanism receives from another, according to the positive rules of a mechanism' (MM 158). To overlook this crucial stratification of mechanisms is to 'mask the pure productive essence, the positional process by which logic, as a machine, never lacks anything that it does not produce elsewhere' (MM 153). It is under this logico-mathematical paradigm that the sciences, whose essence lies in formalization, are envisioned as stratified tectonics of incorporeal machines, whose irreducible stratification forecloses the possibility of suture.[64]

Neither his refutation of Miller, nor his similar critique of the abuses to which Lacan subjects Gödel's incompleteness theorem, exhaust the scope of Badiou's text. If stratification precludes suture, Badiou argues, this is not because the Lacanian concept of suture is essentially misguided; it is due to the singular nature of mathematical—and by extension, scientific—discourse as such. The concept of suture does, indeed, characterize the relation of the subject to certain signifying chains, but only those chains which are not genuinely scientific. Suture, which places the subject in relation to those discourses in which a subject's placement is required, is conceived by Badiou as a specifically *ideological function*. Where suture is foreclosed, ideology cannot take hold, insofar as it is rendered powerless to sustain the 'space of placements' that is proper to it. The epistemological break is therefore seized as formalization insofar as formalization entails the stratification of its mechanisms and therefore a desuturing of discourse, signalling a victory of the force of thought over place.

As always, it is necessary to add that the process of stratification is necessarily an ongoing one, and is ceaselessly carved out against the indistinct choruses of 'indefinitely stratified' and 'ideologically destratified' (and so, (re)sutured) mathematical thought (MM 172). These choruses compose the ideological material on which mathematical production operates, endlessly re-enacting the dialectic of ideological closure and scientific rupture.

This conception has several clear advantages: it clearly situates the role of psychoanalysis with respect to epistemology, reintroduces the function of subjective placement (the operation of suture) into Badiou's analysis of ideology, and provides an intriguing (even if only transitory) basis for his enduring contention that the discourse produced through

64. Badiou's analysis in 'Mark and Lack' is a likely source of Slavoj Žižek's insightful definition of the term 'suture': 'one could', he writes, 'define suture as the structurally necessary *short circuit* between different levels', Slavoj Žižek, '*Da capo senza fine*', in *Contingency, Hegemony, Universality*, London, Verso, 2000, p. 235. Žižek, incidentally, was once both a student and analysand of Jacques-Alain Miller.

epistemological breaks is of genuinely universal address. Taking a position that appears to be utterly opposed to much of his later (post-Althusserian) work, Badiou argues that science is universal insofar as

> *there is no subject of science.* Infinitely stratified, regulating its passages, science is pure space, with neither reverse nor mark nor place for what it excludes.
>
> A foreclosure, but one of nothing, it may be called a psychosis of no subject—and therefore of all: universal by full right, a shared delirium, it is enough to hold oneself within it to no longer be anyone, anonymously dispersed in the hierarchy of orders.
>
> Science is an Outside without a blindspot. (MM 161-2)

Certain problems, however, follow from the text's thoroughgoing confusion of mathematics' momentary existence with its normative ideal. An example of this can be seen in the passage above, which leads us to ask: how could mathematics, as it exists, at this or at any other historical moment, be infinitely stratified? In the example that Badiou reconstructs in this essay, after all, we can count no more than three, and later four, distinct strata—each produced at a distinct (though rationally reconstructed) moment in the development of the mathematical apparatus in question. In different concrete examples, there may be more, or there may be less active strata, but their number will never be other than finite. To say that mathematics is 'infinitely stratified' is clearly to say that mathematics is involved in a process of ceaseless stratification, such that it cannot be contained in any single totalization. Of course, such a totalization—however detotalized the notion it seeks to express—is precisely what is undertaken here. Only in view of such a totalization is it possible to speak of the absolute foreclosure of the ideological subject. In a certain sense, however, this seemingly contradictory picture is emblematic of Badiou's understanding of epistemology itself, insofar as it is directed towards an 'ideological recovery of science' (CM 9), a recovery which, as ideological, involves a totalizing and normative representation of its object.

The time that passes between 'Mark and Lack' and *Being and Event* finds Badiou rejecting the machinic universalism espoused in the former and struggling to articulate a universalism founded in disciplined subjective fidelity.[65] Evidence of this struggle, and the subtle conceptual

65. It would be irresponsible to mention this period of transition without invoking 1982's *Théorie du sujet*, Badiou's most significant and ambitious project to date at the time of its publica-

manoeuvrings that it demanded, can be gleaned from the notes from a series of logic seminars conducted between 1980 and 1983, preserved thanks once again to the efforts of François Nicolas. There, we find an invocation of the Lacanian notion of the real and its relation to mathematical formalization, an invocation which necessitates a mediation between formalism and subjectivity whose exact outline remains indistinct. These texts mark a pronounced change in Badiou's conception of the real and its relation to thought. In the more or less 'structuralist' writings of the late sixties, Badiou tended to identify the real with simple deductive consistency; eschewing the 'paradoxical' dimension of the real foregrounded

---

tion. It is with some regret that I leave it largely out of the discussion; this much neglected text deserves better attention than I can give it here. I will offer only these few remarks. First, we find rebuilt in that text the general opposition between apparatuses of repetition, totalization and placement and vectors of rigorous transformation that, between 1966 and 1968, characterized Badiou's interpretation of the Althusserian opposition between ideology and the epistemological break. There, it is raised to a properly ontological level of abstraction and schematized as the contradiction between the *space of placement* (to which Badiou gives the neologism, '*l'esplace*') and *force*. Employing a schema that he attributes to Hegel's *Science of Logic*, Badiou argues that all things partake of these two dimensions: each thing is both itself (as a force) and its inscription in a space of placement—a thesis abbreviated as $A = AA_p$, which Badiou refers to as 'the matrix of scission'. This thesis, which is taken to be the very kernel of dialectical thought (or one half of dialectics' 'split kernel'—the other half being the matrix of alienation and sublation), is then given two forms, which ultimately differ regarding the element of the matrix to which they give the last word: either placement dominates, without ceasing to be riven by the forces that it places, or else force predominates and effects a transformation of the space of placement itself. The first option gives us the 'structural dialectic', which tends towards idealism; the second gives us the 'historical dialectic', which exhibits a materialist tendency. To these two dialectics correspond two divergent conceptions of 'the real'. On the one hand, we have a 'structural' vision of the real, such as the one propounded by Jacques Lacan, who sees the real as the insistence of a sheer and ineluctable impossibility, and as an evanescent, 'vanishing cause' that haunts the symbolic order. Lacan gives voice to this vision of the real in dozens of curious aphorisms, one of which is: 'The real is the impasse of formalization'. The other, 'historical' vision of the real, grasps the real as the forced passage of a new formalization, a breach of the impossible crystallized as form. This is the vision specific to Badiou's philosophy. It effectively inverts the Lacanian aphorism, and holds that 'formalization is the im-passe of the real', or, 'the place where the real passes in force' (TS 40, 41). Badiou summarizes his relation to Lacan in a nutshell when he writes:

> if, as Lacan says, the real is the impasse of formalization [...] then we must venture that formalization is the impasse of the real. [...] What is needed is a theory of the pass of the real, through a hole in formalization—where the real is not only what is lacking in its place, but that which *passes with force* (TS 43).

All this is quite summary. For an outstanding elucidation of Badiou's *Theory of the Subject* and its place in the entire network of Badiou's writings, the reader can do no better than Bruno Bosteels' two-part essay: 'Alain Badiou's Theory of the Subject: The Recommencement of Dialectical Materialism?'; 'Alain Badiou's Theory of the Subject: The Recommencement of Dialectical Materialism? (Part I)', *Pli: Warwick Journal of Philosophy*, no. 12, 2001, pp. 200-29, and 'Alain Badiou's Theory of the Subject: The Recommencement of Dialectical Materialism? (Part II)', *Pli: Warwick Journal of Philosophy*, no. 13, 2002, pp. 173-208.

by Lacan and his acolytes, Badiou seized it as a concept immanent to mathematical logic—a concept, as we have seen, that is entirely expressed by Emil Post's definition of consistency.

In his later works, however, consistency comes to be seen as an insufficient basis for the relation to the real from which mathematical thought will draw its Platonic dignity. As Badiou observes in the aforementioned seminars,

> great mathematical thought is only secondarily calculative; it is first of all conceptual. [...] Calculation is not the absence of the subject but its absentation. In calculation, the subject is not lacking but comes to be so [*le sujet ne manque pas mais vient à manquer*]. Formalism is mechanizable. Machines, even if they treat of the real, have no relation to the real. Only the subject has such a relation.

Where these meditations eventually lead is towards a singular synthesis between the deindividuating rigours of formalization and the evanescent point in which the subject is convoked in a cancellation of its own identity.[66] They lead towards Badiou's decision, which Bosteels has illustrated in detail, that the real must be thought not only as consistency but also as vanishing cause.[67] This synthesis of formalism and vanishing cause, whose preliminary stages are worked over at length through the seminars recollected in *Theory of the Subject*, and again in the unpublished seminars of the eighties, comes to be realized in the conjunction of *truth-procedure* and *event* that *Being and Event* sets forth. 'Subject', for Badiou, will henceforth designate this singular pairing—the pairing through which subject comes into being as the agent and support of a process of formalization.

Idiom aside, however, there is less distance than might be imagined between the a-subjective, incorporeal machines that populate Badiou's earliest meditations on mathematics and the radically subjective procedures that come to light in his later texts. The negative commonalities of the two figures are easy enough to enumerate—the 'subject' of Badiou's

66. The non-self-identity manifest in evental subjectivization receives a quasi-mathematical expression in the 'matheme of the event' detailed in Meditation 17 of *Being and Event*. There, we find the event defined in such a way that its identity cannot be established by the set-theoretic axiom of extensionality, the only means that Zermelo-Fraenkel set theory has for the determination of identity. This is done by defining the event as a *non-wellfounded* multiplicity, consisting of the elements of its site on the one hand, and itself on the other. Formally, putting X for the site and e_X for the event, we have: $e_X = \{x \in X, e_X\}$.

67. See Bosteels, 'Alain Badiou's Theory of the Subject: The Recommencement of Dialectical Materialism? (Part II)'.

later works is stripped of many of the same attributes that the image of the machine was enlisted to expel: Badiou's subject is not the individual; it is neither egological, psychical, substantial, nor conscious, and to participate in its constitution is no less an anonymous dispersal into the vicissitudes of a procedural becoming than is the mechanical psychosis celebrated above. Badiou's subjects do not precede the procedures in which they are engaged; but they maintain, within these procedures, a dimension of anticipation and transcendence, and therefore a freedom irreducible to the automatism of symbolic mechanisms.

They are, moreover, *without object*; the figure of the object has no more place vis-à-vis the subject than it does in the stratified terrain of a-subjective psychosis exposed in 'Mark and Lack'. This theme is raised recurrently through Badiou's later writings,[68] notably in an essay-adaptation of *Being and Event*'s thirty-fifth meditation, under the title 'On a Finally Objectless Subject',[69] and in a commemorative article on Jean-Paul Sartre, in which he places the objectless structure of his conception of the subject on a list of his departures from the philosopher who, before Althusser, served as Badiou's first master. There, we read:

> I defend a doctrine of the subject *without object*, of the subject as an evanescent point of a procedure originating in an evental supplement without a motive. There is not, in my eyes, an other-being of the subject, unless it is the situation of which a truth is a truth. [...] The true does not speak of the object; it speaks of nothing but itself. And the subject does not speak of the object either, nor of the intention that sights it; it speaks only of the truth, of which it is an evanescent point.[70]

It is this category of truth, or 'generic procedure', that serves to unite the objectless subject with both the epistemological break and the non-objectal ontology of the multiple in which it will be inscribed. Though in many respects this category occupies the same terrain as its Althusserian ancestor, there are ways in which it is significantly broader than the latter: whereas the epistemological break was proper only to science, truth procedures encompass four distinct species of practices. These include

68. Repeatedly, that is, before being retracted—or amended, it's hard to say—in *Logiques de mondes*, where it is written that 'only a logic of the object, as unity of appearing-in-a-world, allows the subjective formalisms to be sustained through their objective dimension: the body...', p. 205. This is a development we'll leave in the dark for now...

69. Alain Badiou, 'On a Finally Objectless Subject', trans. Bruce Fink, in Eduardo Cadava et al. (eds.), *Who Comes After the Subject?* London, Routledge, 1991.

70. Alain Badiou, 'Saissement, dessaisie, fidélité', in *Les Temps modernes*, vol. 531, 1990, p.20.

science, as before, but also love, politics and art—each of these serving, in addition, as 'conditions' informing the philosophical construction of the system through which they are thought—a point I will come back to at the end of this essay. This fourfold multiplication of the field of the epistemological break is of great significance for understanding Badiou's philosophy as a whole, but we will leave it aside for the time being. The reader, if interested, should consult Badiou's *Manifesto for Philosophy* for a further elaboration of his theory of the ways in which these various forms of truths come to condition philosophy.

As we have seen, the category of epistemological break includes within itself the category of that with which such breaks 'break'; it includes the category of *ideology*. It follows that to make any claim as to the existence of a basic continuity between the categories of epistemological break and truth procedure one must be able to locate, in the epistemological break's descendent, that with which it 'breaks'. Such an element is easily located. In a perplexing change of terminology, Badiou gives to it the name, 'knowledge'. Its structure, however, is strictly homologous with that of 'ideology', as construed by Badiou in his Althusserian texts. This (from an Althusserian perspective) unlikely coupling of the name 'knowledge' with the structure of ideology threatens to give rise to much confusion, and was the likely provocation for Slavoj Žižek's (erroneous) remark that:

> the opposition of knowledge (related to the positive order of Being) and truth (related to the Event that springs from the void in the midst of being) seems to reverse the Althusserian opposition of science and ideology: Badiou's 'knowledge' is closer to (a positivist notion of) science, while his description of the Truth-Event bears an uncanny resemblance to Althusserian 'ideological interpellation'.[71]

In his cursory reference to '(a positivist notion of) science', Žižek nevertheless provides us with a means of untangling this mess. Without making any claims for its philological accuracy, we may recall that the 'positivist

71. Slavoj Žižek, *The Ticklish Subject*, London, Verso, 2000, p. 128. Though I will not address it here, we may note that insofar as ideological interpellation concerns the assignment of 'proper places' to individuals cum subjects, it is utterly inimical to the category of truth, which proceeds to disrupt the existing apparatuses of placement 'with force'. Badiou's most thorough treatment of this dimension of truth-bearing subjectivity is to be found in *Theory of the Subject*. Peter Hallward, in *Badiou: A Subject to Truth*, Minneapolis, University of Minnesota Press, 2003, pp. 148-151, has offered another solid rebuttal of Žižek's interpretation by focusing on the role of religion in Badiou's thought and Žižek's criticism.

notion of science' refers, in Badiou's work, to an epistemology arrived at through an abusive generalization of operations proper to formal semantics—conceiving of the mandate of science to be the regulated association of theoretical predications with the relevant subclasses of the experimental domains that would constitute the theory's 'models'.[72] Of course when, in Badiou's '(Re)commencement' essay, such a notion confronts the 'Althusserian opposition of science and ideology', it falls squarely on the side of the latter. If it carries over into *Being and Event*, even if under the rubric of knowledge, we should suspect something closer to a relabelling rather than a revaluation of the Althusserian opposition.

And carry over it does. Knowledge, for the Badiou of 1988, is schematized as a correlation between predicative propositions (codified as set-theoretical formulae restricted to this or that set/situation) and the subsets of the set/situation over which those propositions range. This set-up is not, of course, *identical* to those found in set-theoretical semantics, but in their broad outlines the two bear a striking resemblance. It is no coincidence that Badiou repeats the gesture, made in 'The (Re)commencement', of taking 'positivist epistemology' to exemplify the regime of classification so described, a comparison whose substance changes little despite the terminological shift.[73] The fundamental operation accorded to knowledge in *Being and Event*—the classification of subsets according to their predicative designations—is virtually identical to that

72. We do seem to find such depictions of science in the work of such early positivist texts as Karl Pearson's *Grammar of Science*, London, J. M. Dent & Sons, 1899, where the fundament of science consists in little more than the diligent *classification of facts*. Indeed, Pearson argues, 'The man who classifies facts of any kind whatever, who sees their mutual relations and describes their sequences, is applying the scientific method and is a man of science', p. 16.

73. The comparison in question, as well as the Badiou's general description of 'knowledge', is to be found in Meditation Twenty-Eight of *Being and Event*, from which we quote the following paragraph:

> Positivism considers that presentation is a multiple of *factual* multiples, whose marking is experimental; and that constructible liaisons, grasped by the language of science, which is to say in a precise language, discern laws therein. The use of the word 'law' shows to what point positivism renders science a matter of the state. The hunting down of the indistinct thus has two faces. On the one hand, one must confine oneself to controllable facts: the positivist matches up clues and testimonies, experiments and statistics, in order to guarantee belongings. On the other hand, one must watch over the transparency of the language. A large part of 'false problems' result from imagining the existence of a multiple when the procedure of its construction under the control of language and under the law of facts is either incomplete or incoherent. Under the injunction of constructivist thought, positivism devotes itself to the ill-rewarded but useful tasks of the systematic marking of presented multiples, and the measurable fine-tuning of languages. The positivist is a professional in the maintenance of apparatuses of discernment (BE 292).

accorded to ideology in 'The (Re)commencement'. The comparison is complicated, however, by the fact that Badiou constructs the entire apparatus of knowledge *within* the framework furnished by the Zermelo-Fraenkel axiomatic of set theory and its models—an installation resulting from the aforementioned decision to identify set theory as the science of being. What we have in *Being and Event* is a concentrate of the 1966 identification of ideology and semantic correlation—already found to be untenable by 1968, when *The Concept of Model* was delivered—projected into a specific aspect of the semantic field. The general postulation that the 'reproductive discourse' linking subsets and predicates offers a structural analysis of ideology, is refined by way of the notion of *constructibility*, derived from Gödel's work on the consistency of the continuum hypothesis and the axiom of choice. The relation of constructibility over a given set connects predicative formulae (formulae with one free variable) restricted to that set, to the subsets that they predicate (*i.e.* subsets which are such that membership of x in those subsets is equivalent to verification of their corresponding formulae by the substitution of x for their free variable).

Drawing on Paul Cohen's work on the independence of the continuum hypothesis, Badiou then describes the ontological trajectory of truth procedures as being *generic multiplicities*. Those multiplicities are generic which elude description by any of the predicates belonging to the field of knowledge proper to the situation in question—hence Badiou's frequent description of the generic as 'the indiscernible'.[74] What the concept of the generic brings to the category truth is a new way of specifying its rupture with, or subtraction from, designation and representation. In *Being and Event*, the break with the reproductive discourse of designation and predication is thus conceived as a subtraction that takes place *within* the domain of that very discourse, rather than as a machinic stratification operating *upon* that domain. Consequently, 'knowledge' (ideology's descendent) and the 'truth procedure' (heir to the epistemological break) are two articulations *internal* to the domain of formal (set-theoretic) semantics: one outlines the predicative framework that the other transgresses, though by doing so it anticipates a second model structure (or 'situation') by way of what Cohen terms a 'forcing relation'. It is here that the crucial

74. More formally, a subset a of S is generic if and only if no formula with one free variable and restricted to S, is a necessary and sufficient condition for x's membership in a, when x (an element of S) is substituted for the free variable in said formula. For a detailed presentation, see Paul J. Cohen, *Set Theory and the Continuum Hypothesis*, W. A. Benjamin, 1966, or Badiou's own presentation of the concept in *Being and Event*, Meditation Thirty-Three.

category of *remaking* comes back to the fore, once again coupled with the word '*forcing*', which now serves not only to invoke an image of upheaval, resistance or might but an exact mathematical concept. I will not enter into the details here; suffice it to say that forcing is a relation defined between the (anticipated) elements of the generic set and statements satisfied in the new model structure, whose composition is obtained by an algorithm operating on the generic set, supposedly complete. The new model is, quite precisely, determined to be an extension of the old, in a way akin to (but distinct from) Robinson's 'immersion' of the reals in a non-Archimedian field. By the same token, it is a pervasive remaking of the old, and a transformation of its structure, capable of changing a wide array of structural properties (it may result in making two transfinite cardinals equal where they were previously unequal, or vice-versa, and, most famously, it may result in determining the cardinality of the continuum to be virtually[75] any transfinite cardinal whatsoever!). And, analogous to the inscription of an infinity-point, the forcing operation as defined by Cohen effects the transfiguration of the indiscernible, generic subset into an *element* of the second model structure.

Forcing and genericity, argues Badiou, suffice to describe the truth procedure 'in its being', and, by crystallizing the idea of a multiple assembled without reference to any predicative particularity and the mathematical potency of such a concept, it elegantly grounds the power and universality of truths on the basis of their ontological composition. It is in being faithful to the indifference of truths to predication that the subject of any truth procedures sustains a discourse addressed *to all*.[76] Nevertheless, neither the essential relation between truth procedure and event is clarified by the concept of the generic, nor are the complex phenomenal accidents that constitute truth procedures in their historical actuality. It is, in part, as a way of bridging these gaps that the concept of formalization makes its return. While the place for the concept seems to have been prepared in *Being and Event* by way of a skeletal theory of the 'operator of fidelity', and more significantly by the reinscription of the essential capacities of subtraction and remaking at the heart of a mathematical ontology, the concept returns in its own name only in *The Century*, as we have seen.

75. Due to Konig's lemma, the continuum, the power set of ω_0, cannot be equal to ω_{ω_0}

76. A compelling account of the subjective dimension of truth's genericity can be found in the historical 'case study' that Badiou provides in *Saint Paul and the Foundations of Universalism*, trans. Ray Brassier, Stanford, Stanford University Press, 2003.

VIII. The Structure of Philosophical Intervention, and its Use of Formal Inscription

The analysis of the categories of epistemological break and truth procedure leads us quite naturally to consider the sort of relation that these categories suppose between the philosophical discourse in which they take shape and the extra-philosophical processes at which they aim. Badiou's constant scrutiny of this relation marks what is no doubt one of his most significant fidelities to Althusser, whose conception of philosophy as an intervention in a determinate ideologico-scientific conjuncture profoundly shapes not only the form of the present book, but the entire 'theory of conditions' that develops and reconfigures itself throughout Badiou's work, and which is given its canonical expression in his *Manifesto for Philosophy*. It is with a brief of recollection this dimension of Badiou's thought—which is perhaps his most significant contribution to the general practice of philosophy—that I will end this introductory essay.

It is necessary to insist on the fact that the four generic procedures that animate Badiou's later philosophy—science, art, politics and love—are strictly *external* to philosophy itself, even though it is philosophy that supplies them with the notions that render their 'compossibility' intelligible. This exteriority means that philosophy 'does not itself produce truths', a situation which, Badiou remarks, 'is quite well-known; who can cite a single philosophical statement which one can meaningfully say is "true"?'[77] This theme, which did not escape the notice of the century's positivists,[78] was also an essential thesis of Althusser, who, in his

77. Alain Badiou, *Manifesto for Philosophy*, trans. Norman Madarasz, Albany, State University of New York Press, 1999, p. 35.

78. Wittgenstein's *Tractatus Logico-Philosophicus*, trans. C. K. Ogden, London, Routledge, 1922, if not the founding text of logical positivism itself, is certainly the foundation of the logical positivists' conception of the relation between philosophy and truth. The canonical passage in this regard seems to me to be 4.11-4.111:

> The totality of true propositions is the total natural science (or the totality of natural sciences).
>
> Philosophy is not one of the natural sciences.
>
> (The word 'philosophy' must mean something which stands above or below, but not beside the natural sciences.)
>
> The object of philosophy is the logical clarification of thoughts.
>
> Philosophy is not a theory but an activity.
>
> A philosophical work consists essentially of elucidations.
>
> The result of philosophy is not a number of 'philosophical propositions', but to make propositions clear.
>
> Philosophy should make clear and delimit sharply the thoughts which otherwise are, as it were, opaque and blurred.

introductory lecture to the *Philosophy Course for Scientists*, pointedly tells his audience that, '[n]ot being the object of scientific demonstration or proof, philosophical Theses cannot be said to be "true" (demonstrated or proved as in mathematics or in physics)'.[79] Even so, they 'can be said to be correct [*juste*] or not' (ibid.). What does this 'correctness', which we are not to confuse with 'truth', mean for Althusser? It signifies, to begin with, a particular relation to 'practice', to the philosophical practice of intervention in the contemporary 'conjuncture' (or 'situation', we could say) of science and ideology, an intervention aiming 'to draw a line of demarcation between the ideological of the ideologies on the one hand, and the scientific of the sciences on the other', an act which Althusser nominates as the 'primary function of philosophy' (PSPS 83).[80] This demarcation, however, takes place at the same time as a certain, ideological reuptake of scientific concepts in the assembly and production of 'philosophical categories' (PSPS 81), and cannot be thought apart from this motion. The preceding exegesis of the 'epistemological break' is a case in point, where the tool capable of demarcating science from ideology is, precisely, a synthesis of the former's concepts (in Badiou's case, the concept of formalization) and ideology's notions (the notion of dialectical materialism). The 'correctness' of any such movement of demarcation and appropriation is, for Althusser, decided politically—it is much the same notion of 'correctness' that one finds in Lenin or Mao. What is crucial in such a decision, Althusser argues, is the question of exploitation.

A philosophy is said to 'exploit' science when it seeks to profit from its momentary impasses and subordinate its efficacy to values and methods of philosophy's device, often reinforcing ideological obstacles already endemic to scientific practice. This exploitation can be undone only by a

---

The kinships between Wittgenstein and Althusser, and the 'Circles' (the famous Vienna Circle and the lesser-known Le Cercle d'épistémologie, whose membership included Alain Badiou and Jacques-Alain Miller, among others, and which was responsible for the publication of *Cahiers pour l'analyse*) whose existence their thought provoked, are as pronounced as their utter differences. It would be fascinating to examine these relations in greater detail, and shed light on how, during the briefly 'structuralist' period of French thought, 'young philosophers anxious to break the history of our discipline in two, progressively became positivists of a new sort' Badiou, 'Préface de la nouvelle édition'.

79. Louis Althusser, 'Philosophy and the Spontaneous Philosophy of Scientists', in Gregory Eliot (ed.), *Philosophy and the Spontaneous Philosophy of Scientists and Other Essays*, London, Verso, 1990, p. 74. Henceforth cited PSPS.

80. This is '*Thesis 2*' in Althusser's lecture. It is later strengthened by '*Thesis 22*. All the lines of demarcation traced by philosophy are ultimately modalities of a fundamental line: the line between the scientific and the ideological' (PSPS 99).

'correct' act of demarcation, since science, itself, lacks the capacities necessary to critique the exploitations that philosophy sets upon it. What is needed for this is

> a force of the same nature as the forces that are in contention: a *philosophical* force But not just any philosophical force: a force capable of criticizing and dispelling idealist illusions [...] that is, a *materialist* philosophical force which, instead of exploiting, respects and serves scientific practice. (PSPS 137)

It is precisely such a force that Althusser's circle sought to deploy in the Course of which the present book is a part.

A certain resonance with this project can, to be sure, still be heard in Badiou's later work, in its manifest effort to disintricate truths from the encyclopaedic regimes that would obscure their novelty and genericity, and to adopt, from these truths, the conceptual material that it will forge into the categories necessary for this task—we have focussed entirely, in this essay, on Badiou's 'philosophization' of concepts drawn from mathematics (model and formalism), but this should not lead the reader to neglect what he adopts from art (a Mallarmean poetics of the event), from politics (militant strategies of universal address, and a general maxim of equality), and from love (the idea of an originary Two, and an idea of fidelity).[81] It is largely Badiou's fourfold multiplication of the idea of the 'break' that destines his categories not only towards the work of demarcation, but that of 'compossibilization' as well—the varieties of truth must be capable of being thought *together*, as contemporary, and not only with respect to their autonomy from the given. 'Philosophical concepts', Badiou writes in his *Manifesto*,

> weave a general space in which thought accedes to time, to *its* time, so long as the truth procedures of this time find shelter for their compossibility within it. The appropriate metaphor is thus not of the register of addition, not even of systematic reflection. It is rather of the liberty of movement, of a moving-itself of thought within the articulated element of a state of its conditions. Within philosophy's conceptual medium, local figures as heterogeneous as those of the poem, matheme, political invention and love are related, or may be related to the singularity of time. Philosophy does not pronounce truth but its *conjuncture*, that is, the thinkable conjunction of truths.[82]

81. These are only examples. Several more can no doubt be found.

82. Badiou, *Manifesto for Philosophy*, p. 34.

In order for this conjuncture to take place, the work of purification is set into a permanent tension with the work of philosophical construction. This tension that was certainly latent in Althusser's work, as well as in Badiou's early writings, insofar as demarcation was itself an admittedly ideology-laden operation, but it is raised to the level of a methodological principle in Badiou's later work. This explains, in part, the increased freedom of language one finds in *Being and Event*, where, in contrast to *The Concept of Model*, threads of analogy and rhetoric are granted unprecedented liberty in order to 'weave a general space in which thought accedes to time'. Nowhere is this contrast more evident than in the use of mathematics in the two books, a contrast that frames several of the problems we pose in our interview with Badiou, particularly the one bearing on the ontological status of mathematics.

To give the reader some sense of what may be expected in passing from Badiou's better known later works to the text he or she is about to read (or, impatient with long introductions, has read already), I will end this essay with Badiou's own remarks on the difference in question. Fittingly, they come from his preface for the new French edition of *Le Concept de modèle*:

> With respect to the way in which mathematics are present in this philosophical text, I am struck, rereading this little book of 1969, that what is at stake is a *return* to the logical equipment in such a fashion as to convince the reader that it is *thus* that one must proceed. The didactic is oriented towards a sort of propaganda of formal inscription, taken for the scene where the truth of concepts plays itself out. It is quite close to what Lacan called a matheme. A matheme, as we know, is first of all a formal inscription capable of the integral transmission, without remainder, of a piece of psychoanalytic knowledge [*un savoir psychanalytique*]. Here, I make the formal inscriptions function like 'global' mathemes (those of Lacan are always local, appropriate to *a* conceptual connection) where epistemological concepts can be transmitted through the calculus of signs. This explains the minutiae with which the symbols are introduced and connected. It is *in them* that the efficacy of logical materialism is presented.
>
> Mathematical inscriptions and their concatenation in demonstrations are always present in my principal works of philosophy these days. [...] They are not put to the same use,

> however. We could say that the didactic changes in orientation. I no longer insist that it takes on a mathematized form, or that concepts must be transmitted in the form of mathemes. To the contrary, mathematical inscription and its theoretical context are, rather, points of departure or clarification, which co-present a concept in a formal '*milieu*' different from that of philosophy. In effect, I seek to capture the power of mathematics for the sake of a conceptual development that this capture is *capable* of effecting. In this sense, formalization is not, in my text, what Lacan pretended it was for psychoanalysis: an 'ideal'. It is a source of inspiration and a support, it being understood that, ultimately, the effects of a philosophical text owe their force and duration to the mere arrangement of concepts.

IX. Note on the Translation

This translation was made from a 1972 reprint of the original 1969 edition of *Le Concept de modèle, introduction à une épistémologie matérialiste des mathématiques* published by François Maspero, Paris.

At the time of writing (the Northern hemisphere summer of 2007), some fourteen books authored by Alain Badiou have been translated into English, all of which came to pass in the last eight years. What has emerged from this collective effort is a loosely knit and globally dispersed but on the whole coherent community of translators—and a certain 'voice' that has, through collective artifice, become Badiou's own. This has been the result of both an implicit agreement on terminology (though one not without exceptions, and often developed through uncertain trials and errors) as well as the evolution of a certain style. Both of these factors stand at a certain distance from the present work, however, which shares very little terminology with the later work (so far only books published after the beginning of 1988 have been translated into English), and which exhibits a subtle but undeniable degree of rhetorical and stylistic difference from the later texts. The extremely helpful influence that several other translators of Badiou's work have had on this project can nevertheless not be underestimated, and their work has been instrumental at arriving at the English voice of *The Concept of Model.*

Against this backdrop of what have been, for the most part, osmotically acquired influences, one of the few explicit considerations that went into the stylistic formation of this text concerned its origin as a lecture. Throughout the final rewrites in particular, I strove to keep this dimension of address in view, letting '*on*' be rendered 'we' more often than 'one', allowing a few more contractions through the net, *etc.* I tried to preserve the fluidity of the original whenever this would not require the sacrifice of detail or an undue degree of interpellation on my part.

Where Badiou has quoted external sources in this text, sometimes without explicit citation, I have sought to make use of the standard English translations (and, in some cases, originals) of these texts whenever I could, and provide the requisite citations. This involved adding several footnotes to Badiou's text, and I have marked my own by placing them in bold square brackets [like so].

The terminology used in this text, which was composed with the same ideal of transparency for which Badiou is still well known, offered very few difficulties. A few clarifying remarks are nevertheless in order with respect to individual terms.

I have translated '*cohérence*' and '*incohérence*' by 'consistency' and 'inconsistency' rather than 'coherence' and 'incoherence'. There is nothing peculiar about these translations, especially in a logico-mathematical context where one speaks of 'consistency proofs' but not 'coherence proofs'. Due to some terminological ambiguity, however, some explanation is called for. The terms '*consistance*' and '*inconsistance*', for which there are certainly no better English translations than the widely used 'consistency' and 'inconsistency', have a particular meaning in Badiou's writings on ontology which is not *directly* connected with logical distinction between '*cohérence*' and '*incohérence*'. In Badiou's 'metaontology', 'inconsistency' and 'consistency' are used to refer to different modalities of multiplicity. A consistent multiplicity is one that is seized as *a* multiplicity, as one thing. An inconsistent multiplicity is one that is not considered 'all at once'. Georg Cantor is largely responsible for this acceptation of the terms, but it should be noted that while Cantor's distinction between the inconsistent and consistent is *de jure*, Badiou's is *de facto*, so that any consistent multiple can likewise be conceived in its *in*consistency: 'before' it is 'counted as one', as Badiou says. Badiou thus destroys the symmetry that Cantor established between the two senses of the consistent/inconsistent opposition: for Cantor, to say that a multiple is *inconsistent* is to say that it is *logically inconsistent* (*incohérent*) to think of it as 'one finished thing'. For Badiou, this is not necessarily the case.[83]

The peculiarly French word '*dispositif*' has no obvious equivalent in English. The word 'apparatus' has nevertheless served to translate the term in Foucault's writings, and its meaning has seemed to have broadened in theoretical texts to come closer to the more general and abstract term '*dispositif*'. The relative abstractness of the French term as opposed to its English counterpart need not trouble us here, however. '*Dispositif*' occurs in *The Concept of Model* primarily to designate formal systems, which Badiou delights in comparing to material, even if incorporeal, machines. In this regard, 'apparatus' is a fortuitous translation, and serves well to carry Badiou's materialist and machinic conception of formal systems.

Badiou speaks frequently of logical and mathematical 'experiments' in *The Concept of Model*, and argues explicitly that 'all sciences are experimental' (CM 50). The French term is '*expérience*', which may also be translated as 'experience'. That experiments in particular, rather than experiences in general, are at issue in this book is evident from the context, however, and, with a handful of exceptions, I put 'experiment' and not 'experience' for '*expérience*'.

There is an old taboo in Lacanian circles regarding the translation of 'mathemes', the poetico-mathematical shorthand that the doctor would sometimes employ in his analysis of psychical structures.[84] Lacan, it is said, insisted that his formulae *not* be translated; *e.g.* the letter 'A', which, in numerous formulae, designates the Lacanian concept of the Other [*l'Autre*], should not be rendered in English translations as 'O'. Now, there are two, contradictory, ways to understand this prohibition: (1) mathemes *must not* be translated, and are translated only to the detriment of their formal integrity; (2) mathemes *need not* be translated, and translation should be *discouraged* lest it obscure their strictly *formal* character by suggesting to the reader that translation is, indeed, necessary for such things. It is impossible to accept both, and it is (1) that I am rejecting as false. It is, of course, unnecessary to translate any formalism—and in my view, this can be said more accurately of mathematical logic than Lacan's improvised

83. For Cantor's explanations of the two terms, see Georg Cantor, 'Letter to Dedekind', in van Heijenoort (ed.), *From Frege to Gödel*, p. 114. For an explanation of Badiou's usage, see Alain Badiou, *Being and Event*, Meditation 1. Tzuchien Tho has recently discussed the relation between *consistance* and *cohérence* with Badiou in their interview, 'New Horizons in Mathematics as a Philosophical Condition: An Interview with Alain Badiou', in *Parrhesia*, no. 3, 2007, pp. 1-11.

84. For a clear exegesis of the Lacanian concept of matheme, and its peculiar position between mathematics and literature, I know no better source than Douglas Sadao Aoki, 'Letters from Lacan', in *Paragraph*, vol. 29, no. 3, 2006, pp. 1-20.

notations—just as it is unnecessary to refrain from doing so. All that matters is the legibility of a certain differential *structure*, and not at all the positive characteristics of the terms themselves (which a refusal of translation would presumably seek to preserve). This much I have already sought to explain in the introduction (section IV). As such, I have 'translated' some of the notation employed in *The Concept of Model* for the sake of greater clarity and conformity with both Badiou's later texts and with the standard notational conventions in mathematical logic.[85] The first change I made was to italicize all constants and variables in the logical language. This, I feel, has the general effect of increasing legibility and avoiding confusion between the indefinite article 'a' and the individual constant '*a*'. The second change was to substitute '∀' and '∃' for Badiou's original 'U' and 'E'. That such changes affect nothing of mathematical significance has already been explained, above. In any case, the rest of the notation has remained the same.

This is true even for the two semantic values 'Vri' and 'Fax', which, in earlier drafts of this translation, I rendered 'Tru' and 'Fls', to analogically capture the similarity, mentioned briefly in the French text, between 'Vri' and 'Fax' on the one hand, and '*vrai*' ('true') and '*faux*' ('false') on the other. The reader will probably wonder why I left these unchanged while having no compunction about changing, with seemingly lesser motivation, U into ∀ and E into ∃. My first hesitation came when Oliver Feltham, after reading an early draft of the book, pointed out to me that 'Vri' and 'Fax' are not exactly the 'abbreviations' of '*vrai*' and '*faux*' that they appear to be; phonetically, they bear little resemblance to their etymological origins, and the crude way in which the third letter is hacked off of each word draws attention to a strictly material operation that bears on the letter rather than its sense. They are not abbreviations but eroded material traces. This characteristic is poorly reflected in the pair 'Tru' and 'Fls', which seem more like simple abbreviations of 'true' and 'false'. On Feltham's suggestion, I experimented with various analogues such as 'Tre' and 'Fle', before the irony of trying to translate these terms at all began to outweigh whatever clarity might be gained in doing so. After all, no

85. There is something artificial involved in standardizing Badiou's notation, however, as he himself has not truly held to a single system. In 'Mark and Lack', for example, we find almost the same system employed as we see in *Being and Event*, except that 'horseshoes' rather than 'arrows' are used for conditionals, so that 'If *P* then *Q*', reads '$P \supset Q$' instead of '$P \rightarrow Q$'. In 'Subversion infinitesimale', we find exactly the same system used as in *Being and Event*. In *Number and numbers*, the logical notation conforms to familiar standards, but the empty set is written '0' instead of Ø, and the set braces '{' and '}' are replaced by '(' and ')'.

sooner has Badiou introduced the terms than he writes that

> [o]ne may read these marks, if one wishes, as 'true' and 'false' [*« vrai » et « faux »*]. But this appellation, where we hear the resonance of semantics' intuitive (that is, ideologico-philosophical) origin, is inessential, even parasitic. All that counts here is the permanent impossibility of confounding the two marks, the invariance of the principle of coupling of which they are the inscribed experience. (CM 38)

In the end, I left the terms as I found them: 'Vri' and 'Fax'. This should not obscure the fact that, without causing the least logical damage to the text, I could have translated them 'Dog' and 'Cat'—or even 'Fax' and 'Vri', switching their order. It would have felt silly, and somewhat obnoxious, but could have been done without losing an ounce of rigour.

The Concept of Model

Foreword

The first part of this text (Chapters 1 through 5) reproduces a lecture [*exposé*] given by Alain Badiou on 29 April 1968, within the framework of the Philosophy Course for Scientists held at *l'Ecole Normale Supérieure.*

The rest (Chapters 6 through 10) were to be the object of a second lecture, due to be given on 13 May 1968. That day, as is well known, the popular masses, mobilizing against the bourgeois, Gaullist dictatorship, affirmed their determination across the entire country, and began the process that would lead to a far-reaching confrontation between the classes, turning the political conjuncture on its head and provoking effects whose aftermath was not long in coming.

As one would imagine, in the midst of this tempest, intervention on the philosophical front fell into the background.

Even today, the somewhat 'theoreticist' accents of this text hearken back to a bygone conjuncture. The struggle, even when it is ideological, demands an altogether different style of working and a combativeness both lucid and correct [*juste*]. It is no longer a question of taking aim at a target without striking it.

What one will encounter in this text is thus not only a document and a landmark, but a project that was happily interrupted.

But maybe something else as well: bearing in mind a sense of the relative scope of the crisis' historical significance and the quality of its actors, one might recall that Lenin, in the wake of his defeat in 1905, accorded, for a moment, an exceptional importance to the philosophical struggle against the empirio-criticists.

This is because apparent failures of political practice, erroneous diagnoses of 'relapse' [*reflux*], petit-bourgeois discouragements, always nourish a race of liquidators, idealists and revisionists, which, failing an instantaneous transformation of the world, or even of 'life', console themselves by simply trying to change Marxist-Leninism.[1]

1. [This term receives some explanation in Althusser's introduction to the Course, published

We entertain no illusions: the region in which this work is situated (the doctrine of science) is not only very limited, but quite indirect, but it would be dangerous for us to be mistaken about the meaning of this limitation. We nevertheless believe that it would be useful to call to mind the angle from which the revival of 'Dialectical Materialism', in our eyes and from our point of view, might be pursued or consolidated.

Théorie, December 1968.

as *Philosophy and the Spontaneous Philosophy of Scientists.* Althusser writes that '[n]ot being the object of scientific demonstration or proof, philosophical Theses cannot be said to be 'true' (demonstrated or proved as in mathematics or in physics). They can only be said to be 'correct' [*justes*]. [...] What might 'correct' signify? To give an initial idea: the attribute 'true' implies, above all, a relationship to theory; the attribute 'correct' above all a relationship to practice' pp. 74-5. See Louis Althusser, *Philosophy and the Spontaneous Philosophy of Scientists*, Gregory Elliot (ed.), trans. Warren Montag, London, Verso, 1990.]

1. A Few Preliminaries Concerning Ideology

We are all familiar with descriptions of a certain ideological formation, which partitions the discourse of science according to a presupposed distinction between empirical reality and theoretical form.[1]

We will recall that this distinction organizes an image of science, defining it, by and large, as the formal representation of a given object. In this configuration, the dominant element may be held to be the effective presence of the object. Such a configuration could be called an empiricism. However, it's possible that we might go back to the formal apparatuses [*dispositifs*] in their anteriority, the mathematical code in which the present object comes to be represented, and take these to be dominant. We would then designate the configuration as a formalism.

It's clear enough that empiricism and formalism, here, have no other function than to be the terms of the couple that they form. What constitutes bourgeois epistemology is neither empiricism nor formalism, but the ensemble of notions by which we designate, now, their distinction, and now, their correlation.

This is exactly how logical positivism, the dominant epistemology of Anglo-Saxon countries for more than twenty years, poses the problem of the unity of science. In a canonical 1938 article entitled 'Logical Foundations of the Unity of Science', Rudolf Carnap proceeds as follows:

a) He explicitly poses the constitutive distinction that we are interested in: 'The first distinction which we have to make', he writes, 'is that

1. cf. Louis Althusser, *Cours de philosophie pour scientifiques*, Instalment I, forthcoming. See also, P. Macherey, ibid., Instalment II, forthcoming. [Instalment I has since been published and translated into English as *Philosophy and the Spontaneous Philosophy of Scientists*. Instalment II has not yet been published in any language.]

between *formal science* and *empirical science*'.[2]

b) He attempts to discover the rules of reduction that would allow the terms of one empirical science to be converted into those of another. He shows, for instance, that the terms of biology are convertible into the terms of physics: physics is a 'sufficient reduction basis' for biology. The use of operators of reduction permits Carnap to affirm the unity of the *language* [*langage*] of science, in the sense that a 'physical' language is a reduction basis for the empirical sciences.

c) He poses the problem concerning the relation [*rapport*] between this unique language and the artificial languages of the first group of sciences, the formal sciences. Carnap's entire semantic analysis culminates in this question, whereby the process which began with the distinction between the two types of science is finally wrapped up [*par quoi se boucle la demarche qu'ouvrait la distinction des deux types de la science*].

Notions like empirical science, reducibility, analysis of meaning, etc., and their refined elaboration, articulate the stages of the position and disposition of this initial distinction.

This articulation is both elaborate and quite particular [*spéciale*]. It is not, in its discursive existence, immediately reducible to the generality of a given ideology. Carnap, moreover, explicitly opposes it to other *variants*, to that of the logician, Quine, for example, who, for his part, effaces the distinction between logical and factual truth at the outset. For Quine, in fact, to admit variables into a logical calculation is to produce a law over the constants that these variables take as values. And the constants are fixed only insofar as they are capable of denoting concrete objects. Reciprocally, what empirically 'exists' is nothing other than that which may be assigned to a constant. Finally, as Quine writes, 'to be is to be the value of a variable':[3] the empirical is a dimension of the formal, or vice versa.

And yet the opposition between Carnap and Quine is *internal* to the same problematic. Quine, in effect, defines the particularity of his enterprise (the originality of his project) by justifying the *negation* of a distinction that Carnap, for his part, aims to *reduce*. If Carnap's discourse has this reduction as its essence, all that matters to Quine is the justification

2. [Rudolf Carnap, 'Logical Foundations of the Unity of Science', in Otto Neurath et al. (eds.), *International Encyclopedia of Unified Science*, vol. 1, Chicago, University of Chicago Press, 1955, p. 45.]

3. [W. V. O. Quine, 'On What There Is', in *From a Logical Point of View: 9 Logico-philosophical Essays*, 2nd ed., Cambridge, Harvard University Press, 1964.]

of the claim that there is no need to reduce what can conveniently be denied. The distinction in question—between 'fact' and logical form—is the common motor of the two discourses.

Or, to be precise: the instability and perpetually aborted rebirth of this distinction represent the the power [*la contrainte*] of the lure over ideological discourses, which it deprives of any access to their proper cause. In principle, these characteristics are those of a discursive *agitation*, which infinitely displaces the essentially empty place where the impracticable Science of science must be inscribed.

Here, we must understand that what separates two ideological discourses is not of the same nature as that which separates, for example, science from ideology (an epistemological break), or one science from another. For their rule separation is precisely the ultimate form of the two discourses' *unity*.

We can compare this with musical *variations* on a theme: they are different, but by virtue of a difference that relates them to one another [*les rapporte l'une à l'autre*] as variations *of the same theme*. The (infinite) system of differences between variations is the effect of the (unique) difference between the theme and that which it is not, but which nonetheless relates to it [*s'y rapporte*]: the field of possible variations, the variational space. There are no variations except for those which occur in this space, which no variation justifies, because it is the place where, counterbalancing one another in a unity, differences are established. The ideological lure leads us to attribute, to the variations themselves, the causal power behind the systematic unity of their differences, thereby confusing the *trajectory* of the system with the law of its *production*, with which the singular lack *of theme* must be connected.

It has been shown that to speak of 'Science' [*la science*] is an ideological symptom—as it is, in truth, to speak of ideology in the singular.[4] Science and Ideology are plural. But their types of multiplicity are different: the sciences form a discrete system of articulated differences; the ideologies form a continuous combination of variations. Let's take this assertion as a thesis, and propose the following *definition*:

Given an ideological formation, characterized by a couple of terms, we will call a *variant* any system bound to notions that permit of post-posing [*post-poser*], and possibly answering, the question concerning the unity of the two terms.

I say post-posing, because the unity of the couple is always already a

4. cf. Louis Althusser, *Cours de philosophie pour scientifiques*, Installment I.

precondition of the existence of the ideological discourse under consideration, so that the question of that unity is always a pure and simple repetition. Marx says—almost—that man does not pose any problems that he cannot resolve. Here it is necessary to say: we pose only those questions whose answers are the pre-given conditions of the questions themselves. As such, it is the rule of this repetition to go unnoticed by those who perform it. And this invisibility is developed through the artifice of variants. To return to the musical metaphor: these discourses are the variations on a theme *which is not given* (which does not figure amidst the variations, nor in the head, nor elsewhere), so that, for itself, each variation cannot but be an image, seized in its presence, of the theme in person. And so every variant dogmaticizes over its own priority.

The proliferation of methodologies in the pseudo-sciences that make up the self-proclaimed 'human sciences' reflects the infinity of variational principles, as well as their misunderstanding [*méconnaissance*].

2. On the Theses to be Defended in the Sequel

We will call the units of ideological discourse *notions*, the units of scientific discourse, *concepts*, and those of philosophical discourse, *categories*.

Philosophy being, essentially, the ideological recovery of science, categories denote 'inexistent' objects in which the work of the concept and the repetition of the notion are combined. For example, the Platonic category of 'ideal number' designates, in an 'inexistent' arrangement, concepts of theoretical arithmetic and hierarchical notions of ethico-political origin; the Kantian categories of time and space relate the relative notions of the human faculties to the concepts of Newtonian physics; the Sartrean category of History combines Marxist concepts and ethico-metaphysical notions, such as temporality, or freedom, etc.

That said, we formulate the following theses:

Thesis 1: There exist two epistemological instances of the word 'model'. One is a descriptive notion of scientific activity; the other is a concept of mathematical logic.

Thesis 2: When the second instance serves to support the first, there is an ideological recovery of science, which is to say, a philosophical category, the category of model.

Thesis 3: The philosophical task at hand is to disintricate, from amongst the uses to which the category of model is put, a *subservient* usage, which is nothing but a variant, and a positive usage, invested in the theory of the history of science.

3. On Certain Uses of Models that are Not in Question Here

The first part of Thesis 1 is illustrated perfectly in a well-known methodological text by Lévi-Strauss, situated at the end of his *Structuralist Anthropology*.[1] Here, the empiricism/formalism couple assumes the form of an opposition between the neutral observation of facts and the active production of a model. In other words, science is here conceived as the confrontation between a real object, about which one must inquire (ethnography), and an artificial object whose purpose is to reproduce the real object, imitating it in the law of its effects (ethnology).

Insofar as this object is artificial (or 'constructed', as Lévi-Strauss has it), the model is controllable. One can 'predict how the model will react if one or more of its elements are submitted to certain modifications'.[2] This foresight, in which the theoretical *transparency* of the model resides, is evidently tied to the fact that the model is wholly assembled [*monté*] ('jerry-rigged [*bricolé*]', says Lévi-Strauss) in such a way that the characteristic opacity of the real is absent from it. From this point of view, the model is not a practical transformation of the real (of *its* real): it belongs to the register of pure invention, and is given over to formal 'irreality'.

Thus characterized, models make up a broad class of objects.[3] For ease of exposition, I will partition this class into two groups: 'abstract' models

1. [Claude Lévi-Strauss, 'Social Structure', in *Structural Anthropology*, vol. 1, trans. Claire Jacobson & Brooke Grundfest Schoepf, London, Basic Books, 1963.]

2. [Lévi-Strauss, 'Social Structure', p. 279.]

3. For a series of examples, see: Michel Serres and Alain Badiou, *Modéle et Structure* (the 5th part, above all), transcript of an academic television broadcast. In *Emissions de philosophie pour l'année scolaire* 1967-8, published by l'Institut Pédagogique National. [For whatever reason, there is no record of a fifth part of this broadcast having ever existed. Only three parts are known to have been produced, of which only the second and third are now extant. Tzuchien Tho has recently translated the transcripts of parts two and three into English, which will appear in a forthcoming volume of *Cosmos & History*.]

and material assemblages [*montages*].

The first group consists of what may be called scriptural objects, that is, properly theoretical or mathematical models. In fact, what are at stake are *clusters of hypotheses* that are supposed to be complete relative to the investigated domain, and whose consistency throughout the deductive development are guaranteed by a code [*codage*], which is generally mathematical.

A choice terrain for such models is cosmology. In the book *Cosmologies au XXth siècle*, Jacques Merleau-Ponty systematically studies, without otherwise going beyond a simple chronicle of the science, the *models of the universe*: indeed, the All never being susceptible to experimental inscription, cosmology is bound to the idealism of the model. These deductive constructions are born of a convergence between theoretical developments of relativity, on the one hand, and astronomical experimentation, culminating in the discovery of the red shift of the spectra of nebulae, on the other. The model is a body of statements in virtue of which this historical convergence is integrated in a unique discourse. Naturally, a diversity of integrations exists, and none of them have the force of a law. This is because the models are nothing but intra-scientific constructions. Just as the child comes to surmount the horror of his fragmented body in the dupery of the mirror,[4] models reflect the momentary [*instantané*] disorder of the production of knowledge in accordance with the premature ideal of a unifying text. The model appertains to the securing metatheory of a conjecture.

In the second group, one finds material assemblages, whose objective is threefold:

(1) To spatially present non-spatial processes in a synthetic fashion: graphs, diagrams, etc.

For example, the information furnished by national accounts facilitates the construction of a graph animated by five curves [*sommets*]: administrations, domestic affairs, goods and services, enterprises, market finances. The mobile fluctuations between these curves outline the structure of exchanges, with graph theory permitting certain refinements with respect to the speed and dimension of the fluctuations.

This is an occasion to indicate what, in a general fashion, bourgeois political economy accomplishes through the construction of models of

4. [cf. Jacques Lacan, 'The Mirror Stage as Formative of the *I* Function as Revealed in Psychoanalytic Experience', in *Écrits*, trans. Bruce Fink, New York, W. W. Norton, 2006, pp. 75-81.]

balanced expansion [*expansion équilibrée*]: here again, the model wards off capitalist 'disorder'—not through knowledge of its cause (such knowledge being the Marxist science of social formations and the intellection of the class struggle)—but through the integrated *technical image* of the interests of the bourgeois class. 'Expansion', presented as progressive norm, is in reality the ineluctable effect of structures in which profit is engendered with the asymptotic decline of expenses. 'Balance' is the rule of security against the exacerbation of contradictions, and the political risk of an extreme ascent of the class struggle. The models of balanced expansion, in the guise of thinking their object (the economy of the ostensibly 'industrial societies'), *objectify class objectives*. A national economy in balanced expansion represents the satisfied *motivation* of statist interventions under the name of 'general interest'. As a portable image, the model externally unifies a political economy, legitimates it, and conceals its cause and rule.

It is of foremost importance to show how economic enslavement and the increasing use of ostensibly 'mathematical models' in economics is one of the clearest forms of revisionism, derailing Marxism at the very core of its most finely crafted part and inexorably aligning it with the objectives of the bourgeoisie.

(2) Other models, always in the second group, endeavour to realize formal structures, that is, to transfer scriptural materiality into another 'region' of experimental inscription. The classic book by Cundy and Rollet, *Mathematical Models*, shows, for example, how to effectively construct, in cardboard or wood, the five regular convex polyhedrons, how to fabricate a machine to trace Bernouilli's lemniscate, and even how to produce a logical connective in the form of a simple electrical circuit.[5]

(3) Finally, another class of models aims to imitate behaviours: this is the vast domain of automatons.

Of course, it is out of the question for the epistemologist to deny the existence of these apparatuses [*dispositifs*], or even their 'regulatory' importance in the history of a science such as cosmology, or their techno-political importance in sciences such as industrial automation [*automatique*] and economics.

We are constrained to saying that the model, whether technical instantiation [*moment*] or an ideal figure, takes its place alongside scientific practice at best. We will note that this transitory adjuvant is destined only

5. [H. M. Cundy & A. P. Rollet, *Mathematical Models*, Oxford, Clarendon Press, 1961.]

for its own dismantlement, and that the scientific process, far from securing it, deconstructs it. Bachelard demonstrates quite well how Bohr's 'planetary' model delivered a useful image of the atom only during the time when microphysics emphasized the effacement of the orbits, the blurring of their tracks [*tracé*], and finally the renunciation of the image itself in favour of the statistical model.[6] Whoever did not know how to renounce the model renounced knowledge: every hesitation [*arrêt*] over the model creates an epistemological obstacle. This indicates the degree to which the model remains in the margins of the production of knowledge [*connaissances*]. And yet, ultimately, there can be no question of rejecting it here.

6. Gaston Bachelard, *L'activité de la physique rationaliste*, [Paris, Presses universitaires de France, 1951] chapter II, especially part 7 of this chapter.

4. On a Purely Ideological Use of the Word 'Model'

An epistemological problem surges up against every proposition struggling to describe the difference and the relation [*rapport*] between model and empirical reality; against every enterprise knotting together ways of thinking that which, in the model, speaks of its object; and against every placement, outside the model, of the thing whose model it is.

There is an epistemological problem if we pretend that the invention of models constitutes the very activity of science. That is, if we present scientific *knowledge* as knowledge via models.

Such is the opinion of Lévi-Strauss in the text that I have cited, and which must therefore be questioned anew.

Let us first of all note that, on this point, the expressions that Lévi-Strauss uses are extremely vague. Models, he tells us, are constructed 'after' [*d'aprés*] empirical reality.[1] 'The model', moreover, 'should be constituted so as to account [*rendre compte*] for all the observed facts'.[2] The word 'account' [*rendre compte*] (further on, one finds 'describe' and 'explain') alone supports the epistemological charge.

Now, the 'observed facts' that the model explains [*rendraison*] are in a state of neutralized dispersion: they are given as such, outside of any theoretic intervention, because this intervention begins precisely with the construction of the model, with the *artifice* of assemblage. Lévi-Strauss, in sum, transfers, to epistemological discourse, the institutional opposition of the ethnographer 'in the field', the attentive collector of customs, and the ethnologist in town, the armed commander [*ordonnateur armé*] of his archived people—or even the speculative opposition between Nature (the continuous opacity of what occurs) and Culture (the patchwork

1. [Lévi-Strauss, 'Social Structure', p. 279]

2. [Lévi-Strauss, 'Social Structure', p. 280 (translation modified)].

[*bricolage*] of denumerable differences). Lévi-Strauss thus opposes, in the positivist tradition, passive information to an activity whose meaning is to reproduce the order in which the information assembles itself.

But how does one *control* this reproduction? What are the criteria for a 'good' model?

In an experimentalist conception of science, like Bachelard's conception of physics[3] or Canguilhem's conception of physiology[4], the experimental 'fact' is itself an artefact: it is a material scansion of the proof, and never pre-exists it. Balibar has shown that under these conditions the dialectic of science is thoroughly internal to a process of *production* of knowledge, and that this process is doubly articulated: once according to the *system* of concepts, and again according to the *inscription* of proof.[5]

This conception no doubt opens onto several theoretical problems. For example, one must ask what the efficacious structures of the double articulation might be. What, in the last instance, is the *motor* of science (in the sense in which the class struggle is the motor of history)? Accordingly, these questions call not for a philosophy of knowledge, but a theory of structural causality, which interrogates science as a practical effect and not as a representation.[6]

In the case of the epistemology of models, however, science is divided into productive intervention on one side, and empirical verification or enquiry on the other. The question of this intervention's sense and value is, at this point, inevitable, by the very logic of such an apparatus.

To ask this question is first of all to take stock of the multiplicity of models. The empirical, being inactive, indicates nothing by itself: all trials [*tentatives*] are possible in the inventive freedom of artifice. The model does not, in effect, administer any proof. It is not *constrained* by a demonstrative process, but merely *confronted* with the real. It is conceivable for such a regime that in times of uncertain research models may 'swarm about', as Serres says.[7]

---

3. Gaston, Bachelard, *Le nouvel esprit scientifique*, [10th édition, Paris, Presses universitaires de France, 1968,] Introduction and chapter 6. [*The Formation of the Scientific Mind*, trans. Mary MacAllester Jones, London, Clinamen Press, 2002.]

4. Georges Canguilhem, 'L'expérimentation en biologie animale', in *La connaissance de la vie*, [2e édition, Paris, J. Vrin, 1965.]

5. Etienne Balibar, *Cours de philosophie pour scientifiques*, Instalment II. [Unpublished].

6. For an exposition of these problems: Jacques-Alain Miller, 'L'action de la structure', in *Cahiers pour l'analyse*, no 9, second trimester, 1968.

7. Roger Martin, *Logique contemporaine et formalisation*, [Paris, Presses universitaires de France, 1964,] especially chapter 4.

And so, if the model represents the truth of scientific work, this truth is never anything other than the best model. The dominance of empiricism is thus restored: theoretical activity cannot come to a decision amidst the necessary multiplicity of models, because it is precisely the activity by which the models are fabricated. It is thus the 'facts' that settle things, by designating the best model, which is to say, the best approximation of the facts. 'The best model', Lévi-Strauss writes, 'will always be that which is *true*, that is, the simplest possible model which, while being derived exclusively from the facts under consideration, also makes it possible to account for all of them'.[8]

The circle here is obvious: to the question 'what is a model?' we respond: the artificial object which explains all the empirical facts under consideration. But to the questions What are the criteria of explanation [*du 'rendre raison'*]? What is the true model? the immediate response is: the true model is the one that accounts for all the facts. For good measure, we add the classic condition of elegance: the model must be the simplest.

In these criteria—exhaustiveness and simplicity—we can recognize the norms of classificatory reason of the classical age, and the fundamental categories of a philosophy of representation. They are even the criteria of eighteenth century pictorial criticism, which is not at all surprising. For the epistemology of models, science is not a process of practical transformation of the real, but the fabrication of a plausible image. Thus, of all the types of models that we have mentioned, the most evidently imitative—automata and economic simulations—have an exemplary function in this doctrine. Lévi-Strauss' constant reference in his text is the classic book by Von Neumann and Morgenstern, *Theory of Games and Economic Behaviour*.[9] The properly scientific contribution of this book is certainly considerable. But it is not this that Lévi-Strauss banks on, but the detestable philosophy that rides its coattails. Lévi-Strauss favourably cites passages where a relation as weak as resemblance is explicitly evoked, such as, for example: '[the models] must be similar to reality in those respects which are essential in the investigation at hand'. Or: '[s]imilarity to reality is needed to make the operation significant'.[10]

8. [Lévi-Strauss, 'Social Structure', p. 281.]

9. [Lévi-Strauss cites this text in endnote 3 (p. 316) of 'Social Structures', in *Structural Anthropology*. The following quotation can be found there as well.]

10. [John Von Neumann & Oskar Morgenstern, *Theory of Games and Economic Behaviour*, Princeton, Princeton University Press, 1953, p. 32.]

It is plain to see how external analogy and simulation are called upon here to reduce the initial gap between the inert opacity of the facts and the constructive activity of models. At its limit, this reduction is complete if one can construct a model of the activity of the construction of models. This is the regulatory myth of the epistemology. It clarifies the strange texts where Lévi-Strauss confers on cerebral complexity the dignity of the structure of structures, the ultimate support of 'structurality' itself. Faced with this final object, one undertakes to construct of a model of cerebral functioning: the aspiration of the cyberneticians, for whom the ideology of models is always the spontaneous philosophy. If science is an imitative artifice, the artificial imitation of this artifice is, in effect, Absolute Knowledge.

Let us summarize:

(1) In this first and somewhat crude form, the word 'model' is the operator of a *variant* of vulgar empiricism. The duality of 'fact' and law is reproduced here by the duality of reality and model. The question of this duality's unity thus takes the form of a reproduction, of functional simulation. The idea of total knowledge [*savoir*], in the end, latches itself onto the cybernetic project of imitating cerebral processes.

(2) This variant is blind to its *objective*, but it is here that the political signification of such a discourse is marked. It is:

— To efface the reality of science being a process of production of knowledge [*des connaissances*], a process which in no way opposes the pre-existence of a real to ideal operations, but develops demonstrations and proofs *internal to* an historically specified materiality.

— To obscure the distinction between the *production* of knowledge [*des connaissances*] and the technical *regulation* of a concrete process. Especially in economic 'models', technical servitude to the conditions of production passes for the atemporal necessity of a 'type' of economy, whose beneficial (profitable) constraints are exemplified by the model.

5. The Scientific Concept of Model and the Neo-positivist Doctrine of Science

We now take up the second part of Thesis 1. The word 'model' figures in indisputably scientific contexts, where it does not pretend to designate the province of theoretical practice but one assignable element within a demonstrative consistency: neither notion nor category, but concept.

This is all one branch, no doubt the most lively, of mathematical logic, which is called the theory of models. Here, the inscription of unambiguous theoretical statements takes place at the end of constrained processes. For example:

— A theory is consistent if and only if it has a model (Gödel/Henkin Completeness Theorem).

— A formal theory that admits of an infinite model necessarily admits of a denumerable model (Löwenheim-Skolem Theorem).

— If set theory without the axiom of choice and without the continuum hypothesis admits of a model, the theory obtained by adjoining these two statements admits of one as well (Gödel's theorem), as does the theory obtained by the adjunction of their negation (Cohen's theorem).

What is behind the word 'model' in these statements, and the often highly complex demonstrations where these statements are obtained? Is there any relation whatsoever between its meaning here, and, say, in the aforementioned texts by Lévi-Strauss and Von Neumann?

An initial inspection of the problem would seem to imply an affirmative response to the second question. If logical positivism has been able to propose a doctrine of science constantly propped up by mathematical logic, this is, among other things, because the concept of model allows it to think the relation between a formal system and its 'natural' exterior. Furthermore, one knows well enough that neo-positivist philosophy has played a leading role in the genealogy of mathematical logic. There has been, historically, a dialectical complicity between logical neo-positivism and model theory.

The reason for this is that a classic distinction between two aspects of logic seems to redouble, within scientific discourse, the inaugural couple of formal and empirical science.

(1) A formal or symbolic system is nothing but a game of inscriptions, whose rules are explicit and which foretell every case without ambiguity. Beginning with an initial set of statements (the axioms), one derives theorems according to the rules of deduction. The *sense* of the game is bound to its internal characteristics: the game could not, for example, have any sense (any interest) if *all* statements were theorems: one would not have any need to play, so to speak; every inscription being legitimate, the rules of deduction would serve no purpose. It will therefore be required that there exist at least one statement that is not derivable from the axioms by application of the rules. This is the fundamental property of the system's *consistency* (cf. Appendix). It is a formal exigency, which we will say expresses a *syntactic* norm. The system's set of rules—concerning the ways in which inscriptions can be *formed* (pure grammar) and *deduced* (sequential grammar) effectively define a syntax. Logical positivism wishes to identify the formal dimension of a science with the syntax of its language.

(2) On the other hand, it is clear that the construction of a formal system is not just a gratuitous game. It essentially aims at tracing out the strict deductive structure, the mechanizable aspect, of an existent scientific domain—that is, a theoretical practice whose effects are inscribed in history. To verify that a formal system expresses that structure well, one must bring its statements into a correspondence with the domain of scientific objects under consideration. Naturally, analogies, resemblances, etc., will not be satisfactory. One must define the rules of correspondence. Everything concerning these rules depends on the *semantics* of the system, on its *interpretation*.

The question of meaning is posed differently, this time: to speak of the meaning of the system is to speak of its various interpretations. The fundamental requirement will be the following: that once the rule of semantic correspondence is constructed, every *derivable* statement of the system (every theorem) must be linked to a *true* statement in the domain of interpretation. 'Truth', here, is nothing but a partitioning of scientific statements, accomplished through the *labour* of concepts, into two classes: true statements (demonstrated, or proven, or any other scientifically assignable form of evaluation), and false statements. Semantics aims to show that one can retrospectively organize this partition through purely mechanical, and entirely controllable, procedures, which are brought into play in a formal system.

If one can effectively assign a 'true' statement to every derivable statement, then the domain of interpretation is said to be a *model* for the formal system.

The reciprocal property is stronger: to every true statement of the model corresponds a derivable formula of the system. In this case, the system is said to be *complete* for the model, etc.

And so there is a whole range of semantic properties. Now, suppose that it were possible to study these properties according to the canons of mathematical rigour: we would thereby be presented with a theoretical *concept* of model.

There is a great temptation to export this concept into general epistemology. One would say, for example, that the purely theoretical or mathematical part of physics is its syntax; that the experimental moment, given to concrete interpretations, is therefore equivalent to the algorithms' semantics; that if the theoretical part of science depends on the evaluation of its consistency, experimentation requires the interrogation of concrete models. The experimental apparatuses would simultaneously be the artifices of these models' construction, and the space where the rules of correspondence between formal *calculation* and concrete *measurement* are exercised.

Every scientific choice would then be implicated in either the (experimental) model and rules of correspondence, or the system and rules of syntax.

Carnap has written a book, *Meaning and Necessity*, whose title already reflects the problematic in question in the opposition-correlation of meaning and necessity: semantic constraint of deduction, semantic exactitude of interpretations. Carnap illustrates this problematic with a simple example: if the experiment [*l'expérience*] can be bound to mathematical algorithms, if it is calculable, this is so insofar as phenomena can be measured. Measurement, through which facts become numbers, is here an essential semantic operation. But every result of measurement is expressed in a *rational number* (more precisely, a number that has only a finite number of decimals), because the 'concrete' operations of measure are necessarily finite. Semantics imposes itself on physics only as a field of numbers grounded in the field of rationals. From a syntactic point of view, however, the limitation to the field of rationals entails considerable complications. For example, the 'square root' operator would not have any generality, because a rational number most frequently does not have a rational square root. It would be preferable, therefore, to utilize the field of *real numbers* (whose decimal expansion can be infinite). The adoption of this field as a base for physics, consequently, stems from an exigency of syntactic simplicity. It appears, then, that the opposition between empirical investigation—to speak like Carnap—and mathematical necessity is pertinent, since it can be found in the types of constraint that it exerts on the adopted language.

The unity of this opposition can also be investigated: it appertains to the articulation of syntactic over semantic constraints. In the above example, the experiment can function as a model of the theory because the field of rational numbers is a sub-field of the field of real numbers. Every measurement can therefore be expressed in a formal language (the system of reals), where the rationals are effectively *marked*; and the forms of calculation, the operations, would essentially be conserved, thanks to a certain invariance of the 'species of structure' [*l'espèce de structure*]: both real numbers and rational numbers form fields—sets in which addition, multiplication and their inverses are everywhere defined (except for the 'inverse' operation of 0, of course).

It would indeed seem legitimate to found an epistemology of models on the systematic study of correspondences between syntactic and semantic concepts.

Is this perspective identical to the one that we've criticized through Lévi-Strauss's text? Yes and no.

— Yes, in that it apparently restores the difference between the empirical and the formal, between the observable and the artificial language where the observable comes to be indexed.

— No, and for many reasons.

a) To begin with, it *overturns* the conception in question. For Lévi-Strauss, it is the formal, the jerry-rigged, the artefact, that is the model, relative to a given empirical domain. For positivist semantics, the model is an interpretation of a formal system. It is thus the empirical, the given, that is the model of the syntactic artifice. A sort of reversibility of the word 'model' thus comes into view.

b) But above all, logical positivism's thesis depends explicitly on a science—mathematical logic—where the key distinction between syntax and semantics functions conceptually.

If we say that the model should 'explain' [« *rendre raison* »] all the facts, our assertion does nothing but redouble—*vary* [*varier*]—the fundamental couple of vulgar epistemology. If, however, we speak of the completeness of a formal system, then we designate a property that can be demonstrated or refuted. This is the object of Gödel's most famous *theorems*, which establish the incompleteness of the formal system of arithmetic, being a formal system that admits recursive or 'classical' arithmetic as a *model*. The criteria of the pertinent syntax relative to a given model are not left to the arbitration of resemblances. They are theoretical properties.

The question of knowing [*savoir*] what finally pertains to the *category* of model is entirely at play here, in the difference between Carnap and Lévi-Strauss—a difference bearing on the exact epistemological relevance of the logical, scientific *concept* of model, which alone can validate or invalidate its exportation for the construction of a philosophical category. We cannot avoid a purely logical detour here.

This detour demands a certain amount of attention, so it's only fair to indicate its objective in advance, thereby emphasizing its necessity: it is a question of placing the (scientific) construction of a concept within the epistemological clearing [*l'eclairage*]. By putting this construction into practice, we first of all hope to lay hold of the exact difference between the concept of model and the homonymous (ideological) notion. Moreover, through the accompanying commentaries, and through emphasizing the deployment of its successive moments, the demonstrative construction will serve to validate another distinction: one which distinguishes two categorial (philosophical) usages of the word 'model'. To put it another way, our reading of science here commands, first [*en amont*] a distance from ideology, and second [*en aval*], a line of demarcation within philosophical discourse: there are two antagonistic styles of discourse *on* science; two forms of ideological re-appropriation of science; and finally, two *politics* of science, one progressive and one reactionary.

I therefore ask the reader not to skip over the technical explications in order to finish quickly. The materialist-epistemological *reality* of what I am trying to introduce is of a piece with an effective scientific practice. Being a matter of mathematical logic, this practice requires scarcely any technical preparation.

6. Construction of the Concept of Model

I. Syntactic preliminaries

At the risk, inherent to the epistemological enterprise, of saying too much for those who practice the science in view, and too little for others, I will offer, by way of example, a stepwise definition of models relative to a very simple but frequently used logical language. The purpose of this is to be elementary in the strict sense, presupposing *no* particular knowledge. I will not be overly meticulous, desiring only to let the articulation of a concept's construction be grasped. For a more extensive development, but one equally attentive to epistemological problems, the reader is referred to Chapter 8, and, for a rigorous treatment, to Chapter 9. It will be useful to keep the exposition that is situated at the end of the text in sight.

We will begin by occupying ourselves with syntax.

Our language of calculation—our game of inscriptions—aims at being an experimental mathematical apparatus [*dispositif*], that is, a system of inscriptions which obeys specific conditions. We must therefore deploy a stock of marks sufficient for the deployment of several 'kinds' of inscriptions, which are the *pieces* of the game.

A) We need to designate the *fixed* differences between our objects; 'object', here, signifies nothing other than what is enchained to scriptural experimentation. We will utilize, for this task a finite or infinite—but denumerable—list of letters: *a, b, c, a', b', c'* ... We will call them the *individual constants*. Note that, as a general rule, they will not be interchangeable in a given inscription.

B) We need to designate the properties of the objects, to mark, that is, certain classes of constants, those which 'satisfy' a property. We will utilize predicative marks, or predicates: *P, Q, R, P', Q'*, ... The simplicity of our example lies in the fact that we admit only 'unary' predicates, capable of marking only one constant at a time. In most mathematical syntaxes, there exist binary predicates, or relations, which mark couples of constants, and even '*n*-ary' predicates which mark a system of *n* constants. The general form of the construction of the concept of model is, nevertheless, essentially the same.

C) Finally, we need to designate the 'generality' of the objective domain, which is to say, any undetermined constant whatsoever, *a place* where any constant at all can come to be inscribed. These undetermined marks will thus eventually be able to be replaced by constants; for this reason they will be called *individual variables*. We will mark them: *x, y, z, x', y'* ... .

We can already *form* certain expressions, or sequences of marks. Not all sequences will be correct: the criterion of syntactic sense—that the game not be completely arbitrary—intervenes here by way of *rules of formation*. We will avoid going into detail. It's clear that the inscription of a constant (or a variable) will be governed by a predicate. For this, providing *punctuation marks* will prove convenient: parentheses and square brackets. For example, $P(a)$ will be a correct (well-formed) expression, which will be read '*a* possesses the property *P*'. Likewise for $P(x)$. Inscriptions of this type, which, punctuation aside, consist of only two marks, will be called *elementary formulae*.

The use of variables is of genuine interest only insofar as it enables us to write general statements, the semantic interpretation of which will be: 'there exists at least one constant marked by the predicate *P*', or 'all constants are marked by *P*'. For this, the classical *quantifiers* are introduced: universal, which we will note ∀, and which is read 'for all' or 'for every', and existential, which we will note ∃ and which is read 'there exists'. A rule of formation authorizes inscriptions of the type:

— $(\exists x)P(x)$, which is read: 'there exists an *x* such that $P(x)$'.
— $(\forall x)P(x)$, which is read: 'for every x, $P(x)$'.

Note that these statements are given here only as examples of acceptable, well-formed, inscriptions and not as 'theorems' or 'true statements'.

In these expressions the quantified variable *x* cannot be replaced by a constant. This is clear enough: the statement $(\exists x)P(x)$ does not tell us *which* particular constant is marked by P, but only that some such constant exists. The statement $(\forall x)P(x)$ tells us that every constant is marked by P, not this one or that. Hence a distinction relative to the type of inscription, highly important in what follows:

Definition: a variable that falls within the scope of a quantifier is called a *bound* variable; otherwise it is called *free.*

Let's take one step further in the combinatory complexity of our apparatus. We wish to be able to construct inscriptions which combine not only letters, but elementary formulae and elementary quantified formulae, and which thereby combine these combinations. For this, we introduce logical operators: the *connectives,* which take 'already' constructed formulae for their arguments. We utilize two here: *negation,* which we will indicate by ~, and *implication,* →. The rules of formation associated with these signs are quite simple:

— If A is a well-formed expression, then so is $\sim A$.
— If A and B are well-formed expressions, then so is $(A \rightarrow B)$.

The first expression is read 'not A'; the second, 'A implies B'.

Finally, it is acceptable to *quantify* the well-formed expressions thus obtained, under the condition that the variable over which the quantifier operates occur freely within these expressions. If, for example, the variable x is *free* in A and in B (if it is not *already* quantified in A or in B), the expression $(\forall x)(A \rightarrow B)$ is well-formed. We are now in a position to write complex well-formed expressions, which we will call the *formulae* of the system. To give an example, and assemble our conventions:

$$(\forall x)[\sim P(x) \rightarrow (Q(y) \rightarrow P(a))]$$

is a formula, which reads 'for every x, if x does not have the property P, then the fact that y has the property Q implies that a possesses the property P'. In this formula, the variable x is *bound* and the variable y is *free.* Such a formula (which contains at least one free variable) is said to be *open.*

$$(\exists x)[P(x) \rightarrow \sim Q(x)]$$

which reads 'there exists an x such that if x has the property P, then it does not have the property Q', is a formula without any free variables: it is a *closed* formula.

It remains for us to give this game its *deductive* form, and present an apparatus that distinguishes, amongst well-formed expressions, those which are theorems (those which can be deduced), and those which are not.

For this, we first define the *rules of deduction*, which permit us to *produce* a formula on the basis of others, through explicit manipulations. Notice that the formulae arranged in this fashion are all well-formed.

In our example, the rules are the following:

1) Given an already-produced expression (or axiom) A, in which the variable x is free, we can 'produce' the expression $(\forall x)$A.

The schema of deduction is thus written (the sign $\vdash$ indicating that the formula A has 'previously' been proven in the system, or that it is an axiom):

$$\frac{\vdash \mathrm{A}\ (\text{with } x \text{ free in } A)}{\vdash (\forall x)\, A}$$

This is called the rule of *generalization.*

2) Given two formulae $(A \rightarrow B)$ and A, we admit a rule of deduction that allows us to inscribe the formula B in this sequence:

$$\frac{\begin{array}{l}\vdash A \rightarrow B \\ \vdash A\end{array}}{\vdash \quad B}$$

This is called the rule of *separation.*

The appendix will convince the reader of the possibilities that these two rules alone offer to the deductive game.

In passing, we insist on the importance of the *effective*, or mechanical, character of these rules (as well as the rules of formation). The philosophical category of effective procedure—of that which is explicitly calculable, by a series of unambiguous scriptural manipulations—is truly at the centre of every epistemology of mathematics. This results from mathematics' properly experimental [*experimental*] aspect, in which this category is concentrated: *the materiality of marks*, the assemblage of inscriptions. Bachelard notes that, in physics, the true principle of identity is that of the identity of scientific instruments.[1] In the investigation of the calculable, and the interrogation into the essence of algorithms, corresponds the

1. Gaston Bachelard, *L'activité de la physique rationaliste*, [Paris, Presses universitaires de France, 1951], Chapter II, especially part 7 of this chapter .

principle of the invariance of inscriptions, and the control of this invariance. The mathematical demonstration is *tested* [*s'éprouve*] via the explicit rule of marks. In mathematics, the moment of verification is represented by inscription.

Once the rules of deduction have been instituted, the initial formulae—the axioms—must be selected. This choice characterizes the theory in question and signals its particularity, since all the other rules of our language (formation and deduction) are general. The choice of axioms makes the demonstrative difference.

We will now set out a concept of *deduction*.

Definition: a finite series of formulae is a deduction if each of the formulae of which it is composed is either:
— an axiom; or
— the result of an application of a rule of deduction to the formulae preceding it in the series.

Every formula (axiomatic or produced) that figures in a deduction is a *theorem* of the system.

Suppose, for example, that we have chosen two axioms:

Ax 1: $\vdash P(x)$
Ax 2: $\vdash (\forall x)\, P(x) \rightarrow \sim Q(a)$

The reader may (without much trouble) verify that the series:

$$\vdash P(x)$$
$$\vdash (\forall x)\, P(x)$$
$$\vdash (\forall x)\, P(x) \rightarrow \sim Q(a)$$
$$\vdash \sim Q(a)$$

is a deduction, following the two rules introduced above (generalization and separation). The formula $\sim Q(a)$ is thus a theorem of the system specified by the two axioms.

It is possible to distinguish between *logical* and *mathematical* axioms. The first, in the scriptural form that characterizes them, do not take any fixed constants into consideration; the second, by contrast, frequently regulate the usage of such constants, which may be called the non-logical symbols of the theory.

In fact, it is frequently the case that we utilize, as logical axioms, infinite series of formulae whose structure (their law of formation, or inscription) is the same. It is thus that *all* (the infinitely numerous) statements of the type $[A \rightarrow (B \rightarrow A)]$, where A and B are any well-formed expressions whatsoever, are often held as axioms in calculi similar to our example. Clearly, constants figure in the majority of expressions of this type. Hence, the expression:

$$[P(a) \rightarrow [Q(b) \rightarrow P(a)]]$$

contains four constants: two individual and two predicative. It is, however, of the type required $[A \rightarrow (B \rightarrow A)]$, and thus figures in the list of axioms. But the constants a, b, P, Q do not characterize this type in the least, nor are they the reason why this formula belongs to the list. The reason for its membership lies solely in the global conformity of the inscription's 'structure' (to the axiom schema). Moreover, in replacing all of the constants by others, or by variables, I obtain a formula that is also in the list, and which is an axiom of the same species. We can therefore see that the axiomatic schema that organizes the list, depending solely on the logical connective that figures therein (implication), is a *logical* schema.

In contrast, consider the following axiom, where S is a fixed predicate and a is a constant:

$$(\exists x)\ [S(x) \rightarrow \sim S(a)]$$

It is clear that the predicate S is altogether particular, and is not replaceable by any predicate whatsoever, no more than is the individual constant a. The axiom (implicitly) defines S as a predicate which possesses *differential* powers of marking with respect to the constant a. The axiom, in effect, poses that there exists at least one constant such that if it is marked by S then a is not. There is an S-incompatibility between a and another (indeterminate) constant. Such a (separative) axiom may be considered as *mathematical*, insofar as it is bound to the experimental apparatus of a particular mathematical theory.

However, a bit further on we will see that the intra-syntactic difference between logical and mathematical axioms is fully thinkable only with reference to the models in which such axioms are 'true'.

7. Construction of the Concept of Model

II. Fundamental aspects of semantics

Here, we will try to make an interpretation 'correspond' to the system, whose syntax we have just described.

The first step is to determine the domain of objects which the correspondence with the marks of the system will be founded. And yet nothing is more indistinct, and more empiricist, than the notion of a collection of objects, to the point that if it maintains this notion, semantics will have no chance of articulating itself scientifically: *it is only to the extent that it deploys the mathematical concept of set, and consequently transforms the notion of domanial multiplicity that the theory of interpretations of a formal system escapes this impotence.*

The name *structure* will be given to the following apparatus:

A) *A non-empty set* V, which we will call the domain, or universe.

Being an 'object' of the structure will mean belonging to this set. And belonging, here, is nothing other than the fundamental sign of set theory, $\in$, and its rigour is that of the theory itself. It already appears that semantics is only a science (and model, a concept) insofar as it establishes itself *within* an existent branch of mathematics, so that the law of the interpretations of a formal (mathematical) system is itself written in (non-formal) mathematics. That we have neither a circle nor absolute knowledge here is what we will clarify in the sequel.

We will utilize the letters u, v, w, u', v'... to mark the distinctions in the universe. We will designate the property of being an 'object' of the universe $u \in V$, and emphasize, in passing, that the production of such an object calls only for an inscription that is *different* from all those which figure in the syntactic apparatus; and so it is true that mathematical experimentation has no material place other than where difference between marks is manifested.

B) *A family of subsets of* V, which we will designate [pV], [qV], [rV] ..., and among which may figure the empty set (the set which has no elements).

Do we have the right to consider such a 'family' as a set, and on that basis assign to it the conceptual rigour inherent in the mathematics of sets? Yes, insofar as this mathematics posits (through the power set axiom) the existence of the set of all the subsets of the given set **V**, of which our family is a definite part. Yes, again, to the extent that this theory axiomatically posits the existence of the empty set.

C) *Two supplementary marks, Vri and Fax.*

One may read these marks, if one wishes, as 'true' and 'false' [*« vrai » et « faux »*]. But this appellation, where we hear the resonance of semantics' intuitive (that is, ideologico-philosophical) origin, is inessential, even parasitic. All that counts here is the permanent impossibility of confounding the two marks, the invariance of the principle of coupling of which they are the inscribed experience.

Every apparatus of the type prescribed by our conditions (A), (B), (C) is a structure. It is in order to bind a formal system to these structures that semantics is employed.

We will suppose that there exists a *function*, designated f, which will be a function of correspondence defined over the syntactic marks such that:

1°) every *individual constant of the system* is made to correspond to an object of the structure. Hence $f(a) = u$.

2°) every *predicative constant* is made to correspond to a subset of the family that defines the structure: $f(P) = [\text{pV}]$.

Note that f operates 'between' the marks of the formal system and those of the structure, transporting the individual constant / predicative constant hierarchy onto another: mark of an element of the universe / mark of a set of elements of the universe.

This transference does not depend upon the simplicity of our example: if the system admitted, beyond its predicative constants, constants for binary relations, being marks assigned to couples of constants, one could take more complex structures into consideration, bringing sets of couples of elements of the universe into play. Set theory, via the axiom of pairs, guarantees the existence of a set whose elements are any two given sets.

The idea that now comes to organize the construction of the concept of model is the following: utilizing the set-theoretic resources of the structure, and the function f, we will give a meaning to the *validity*, or non-validity, for a structure, of a well-formed expression of a formal system. If we are able to establish a relation between syntactic *deducibility* (the fact that the expression A is a theorem) and semantic *validity* (the fact that A is valid for a structure, or for a certain type of structure, or even any structure whatsoever), we may hope to discern the conditions under which a particular structure is a model for a system.

The evaluation of a formula A proceeds step by step, thanks to the marks, Vri and Fax.

To begin, we will pose the following:

Rule 1: $P(a)$ = Vri if and only if $f(a) \in f(P)$;
Otherwise, $P(a)$ = Fax. In other words, the mark Vri is made to correspond to the expression inscribing a's possession of the property P, if the element u which, through f, corresponds to the constant a, belongs to the subset [pV] corresponding to the predicate P.

Rule 2: $\sim A$ = Vri if and only if A = Fax. Otherwise $\sim A$ = Fax. This is the classical interpretation of negation.

Rule 3: $(A \rightarrow B)$ = Fax if and only if A = Vri and B = Fax. Otherwise, $(A \rightarrow B)$ = Vri.

An implication is not 'false' unless the antecedent is true and the consequent false.

We now come to the quantifiers. Let B be an expression in which the variable x is free. We write the expression obtained by replacing, in B, every occurrence of the variable x by the constant a, as $B(a/x)$. We then pose:

Rule 4: Let B be an expression containing no free variable other than x. Then, $(\exists x)B$ = Vri if and only if there exists at least one constant, say a, such that $B(a/x)$ = Vri. Otherwise, $(\exists x)B$ = Fax.

Rule 5: Under the same conditions, $(\forall x)B$ = Vri, if and only if for *all* constants a, b, c, etc., we have $B(a/x)$ = Vri, $B(b/x)$ = Vri, etc.

There remains the case of elementary formulae of the type $P(a)$, and, more generally, the case of open formulae (those which involve non-quantified variables). In effect, our rules permit us to evaluate, step by step, only closed formulae. This is quite normal: the 'truth' of an open formula is not fixed: it depends on the constant that is substituted for the variable. Hence the expression $P(a) \rightarrow P(x)$, where the variable x is free, is, for most structures, false if one replaces x by a constant other than a. In contrast, the expression $P(a) \rightarrow P(a)$ is true for any structure whatsoever. The evaluation of an open formula should therefore take into account all possible substitutions: one must try all the combinations that can be obtained by replacing its free variables with each constant of the system.

We therefore generalize the procedure used for the evaluation of quantified expressions. Let A be an open formula, and let x, y, z... be the *different* free variables that it contains. We will call a *closed instance* of A, a formula of the type $A(a/x)(b/y)(c/z)$, where all the free variables of A have been replaced by constants. Naturally, the number of instances for a given open formula is quite large. This number depends both on the number of different free variables in the formula, and the number of individual constants of the formal system in question. Clearly, all of these instances are closed formulae (without free variables). They can therefore be evaluated by repeatedly employing the five preceding rules.

We will now pose the following, crucial definition:

Definition: A formula A of the system is *valid* for a structure if, relative to this structure, we have A'= Vri for *every* closed instance A' of A.

In particular, a closed formula A is valid if A = Vri, because it has no closed instances other than itself (nothing in it is replaceable).

Note that this procedure is constructed by recursion over the 'length' of the formulae, that is to say, over the *number of marks* that they contain. We begin with elementary formulae of the type $P(a)$, evaluated directly in the structure by examining whether the semantic 'representative' of a belongs to the subset of the universe that represents P. We then follow the procedure permitting the evaluation of A on the basis of the evaluations of the shorter expressions contained within A (or its closed instances), which we assume have already taken place. The evaluation of $\sim B$, for instance, is made on the basis of that of B; that of $(\exists x)B$ on the basis of that of $B(a/x)$, etc.

The conviction that these rules guarantee the existence of an evaluation for any length whatsoever depends on the admission of reasoning by recursion over the integers (here, over the number of symbols entering into the composition of a formula). This suggests two epistemological statements:

1) The rigorous construction of the concept of model, of which evaluation is a moment, implies that the formalized writing be 'numberable' by the natural integers; in other words, that a well-formed expression of the formal system be a *denumerable*, even, for most systems, a *finite* series of indecomposable marks. To speak of a model is to exclude the possibility of a formal language being continuous.

2) After the explicit appeal to the mathematics of sets, we have here an appeal, more or less implicit, to the mathematics of integers, namely to the axiom of induction that characterizes it. To speak of a model is to presuppose the 'truth' (the existence) of these mathematical practices. We establish ourselves within science from the start. We do not reconstitute it from scratch. We do not found it.

Let's go a step further, and state that the rules of deduction of the formal system 'conserve' validity: If A is valid, and if B is produced through the application of a rule to A, then B is valid in any structure where validity is defined. It stands to reason that in reality, the rules are chosen precisely in order to assure a sort of semantic regularity.

Let's quickly verify this assertion for our two rules on page 26, beginning with the schema of generalization. Suppose that A is valid, and that $(\forall x)\, A$ is not. The second part of this hypothesis implies, after the definition of validity, that there exists a closed instance $(\forall x)\, A'$ of $(\forall x)\, A$ such that $(\forall x)\, A' =$ Fax. After Rule 5, this amounts to saying that there is at least one constant a for which $A'(a/x) =$ Fax. But $A'(a/x)$ is a closed instance of A. Now, we have supposed that A is valid; every closed instance of A is thus equal to Vri. This is a contradiction, and our hypothesis must be rejected.

We will note, in passing, that in invoking the principle of non-contradiction to reach our conclusion, we utilize a logic 'in the practical state' [*à l'état pratique*]. It follows that the presupposed mathematics of our conceptual construction (set theory, number theory) likewise mobilize an underlying logic, practical procedures of concatenation, where these mathematical fragments articulate themselves. It is not the case that such 'logical principles' overarch thought (as is the case with the principle of non-contradiction in the metaphysics of Aristotle). These 'principles', on the contrary, form part of what we experience [*expérimentons*] in the field of concrete mathematical production, and have no other existence. For that matter, they have the same status as mathematical statements, susceptible to syntactic verification, within the framework of the assemblage [*montage*] of logical systems.

Now for the rule of separation. To simplify things, we will suppose that all the formulae are closed. If we had B = Fax, Rule 3 would pose that A = Vri entails $(A \rightarrow B)$ = Fax. But we suppose A and $(A \rightarrow B)$ to be valid. It is therefore impossible that we should have B = Fax. B is therefore valid.

Our rules of deduction therefore transfer [*transportent*] validity. From this follows the major consequence that *if a theory's axioms are valid, then so are each of its theorems.* A deduction (cf. page 26) effectively begins with an axiom, and thenceforth involves only axioms or formulae produced through the application of the rules to the formulae which precede them: If the axioms are valid, then so is every formula figuring in a deduction.

The function of correspondence, which supports the procedures of evaluation, thus defines a sort of inference, via the syntactic concept of a deducible statement, from the semantic concept of a statement valid-for-a-structure. We have attained our goal, and we pose that:

A STRUCTURE IS A MODEL OF A FORMAL THEORY IF ALL THE AXIOMS OF THAT THEORY ARE VALID FOR THAT STRUCTURE.

8. Construction of the Concept of Model

III. Some games with the example

I have already evoked mathematics' departure from logic. The surest criterion amounts to saying that an axiom is *logical* if it is valid for *every* structure, and mathematical otherwise. A mathematical axiom, valid only in particular structures, marks its formal identity by excluding others through semantic force. Logic, reflected semantically, is the system of the structural as such; mathematics, as Bourbaki says, is the theory of *species of structure*.[1]

In our example, do there truly exist correct expressions valid for every structure? Certainly. We have mentioned the schema: $A \rightarrow (B \rightarrow A)$, where A and B are any expressions whatsoever. A formula conforming to this schema is always valid, whatever may be the evaluations of A and B, and, hence, whatever the structure. In effect:

Suppose that	$[A \rightarrow (B \rightarrow A)] = \text{Fax}$	(1)
Then (Rule 3)	$A = \text{Vri}$	(2)
And (idem)	$(B \rightarrow A) = \text{Fax}$	(3)
(3), by Rule 3, entails in its turn	$A = \text{Fax}$	(4)

(4) contradicts (2): Our hypothesis must be rejected and we always have:

$$[A \rightarrow (B \rightarrow A)] = \text{Vri}$$

By an abuse of language, we may say: the schema is always valid.

The reader may easily show, for instance, that the schemata

1. cf. the construction of the concept of a species of structure in Nicholas Bourbaki, *Théorie des ensembles*, [Paris, Hermann, 1968,] Chapter 4, *N*° 1. [*Theory of Sets*, New York, Springer, 2004.]

— $(\sim A \rightarrow \sim B) \rightarrow (B \rightarrow A)$
— $[A \rightarrow (B \rightarrow C)] \rightarrow [(A \rightarrow B) \rightarrow (A \rightarrow C)]$

are valid independently of all particularity of structure. These statements are purely logical. Added to the schema above, they suffice, moreover, to define an important logical system: the propositional calculus (cf. Appendix). There exist, evidently, an infinity of other formulae that are purely logical, notwithstanding the fact that they can be inferred from the first three by the rules of deduction, which conserve validity.

The introduction of the quantifiers does not at all deprive certain statements of logical purity, even though 'there exists' and 'for all' depend strictly, with respect to their validity, on the chosen universe. Again, we provide a very simple example. Take the well-formed expression:

$$\sim (\exists x)P(x) \rightarrow [(\exists x) \sim P(x)]$$

which links existence and negation through the predicate P.

Suppose that its evaluation yields the mark Fax. Then (by Rule 3) the antecedent is true and the consequent false. And so:

1st part of the hypothesis:	$\sim(\exists x)P(x) = \text{Vri}$	(1)
which yields (Rule 2):	$(\exists x)P(x) = \text{Fax}$	(2)
which yields, for all constants a (Rule 4):	$P(a) = \text{Fax}$	(3)
2nd part of the hypothesis:	$(\exists x) \sim P(x) = \text{Fax}$	(4)
which yields, for all constants a (Rule 4):	$\sim P(a) = \text{Fax}$	(5)
which yields (Rule 2):	$P(a) = \text{Vri}$	(6)

The result (6) contradicts (3): our hypothesis must be rejected, and the expression in question cannot in any case take the value Fax. It is therefore valid for every structure: it is purely logical.

If we take up this semantic definition of logical axioms, we will see that they say *nothing* about the structures in which the formal system can be interpreted.

Such is the experimental result, so far as the presumed 'transhistoricity' of logic goes. Now, we have already said that there is no contradiction between the logical *practice* inherent in every demonstration, and the construction of particular logical *systems*. Or rather: this contradiction is nothing other than the living dialectic of (semantic) demonstration and (syntactic) experimentation.

In order to establish the 'transhistoricity' of logic, one often argues from what seems like a vicious circle: one cannot have any rational discourse *about* logical principles (except to state their 'evidence'), because rationality is precisely defined by the conformity of a discourse to these principles. Logic would always already be there, consequently conditioning, and not resulting from, the history of Reason.

We are tempted to say that in reality, logic is itself an historical construction, doubly articulated in active principles of concrete demonstrations, and explicit figures of formal assemblage. The 'circle' is resolved in the gap between demonstrative practice and experimental (or 'formal') inscription, the gap that is the *motor* of this science's history. This mode of historical existence does not at all differentiate logic from mathematics.

Finally, the 'transhistoricity' of logic is reduced to the experimental property that *a purely logical system* (*one whose axioms are all logical*) *bears no marking* [*marquage*] *of its models*. Or more precisely: every structure being a model for this system, the concept of model is not logically discernible from that of structure.

Only mathematical axioms lift this semantic indistinction, and produce the effective inscription of a structural *gap*, by which the concept of model is legitimated. Hence a logician such as Church prefers to call the initial, not-purely-logical formulae '*postulates*'.

However, the concept of logic is precisely constructed in accordance with the couple that it forms with the concept of mathematics: it does not encompass [*surplombe*] it. The opposition between mathematics and logic syntactically redoubles the semantic distinction between model and structure. Hence, if, given a particular formal system the difference between two structures is marked in such a way that one is a model of that system and one is not, then it is possible to classify the axioms of that system as either purely logical or mathematical. The former mark a unity where the latter mark a difference.

And yet, the instrument of this conceptual distinction—the concept of structure and therefore set theory—is, itself, mathematical, insofar as this theory, which we take to be formalized, clearly does not admit all structures as models. We will come back to the historical effects of this entanglement.

To conclude, let us give an elementary example of a strictly mathematical statement.

Consider the formula:

$$(\exists x)(\exists y) \sim [(P(x) \rightarrow \sim P(y)) \rightarrow (\sim (\sim P(y) \rightarrow P(x)))]$$

Such a formula cannot be valid for a structure *whose universe consists of only one element*. Suppose, in fact, that for a structure of this type, we had:

$$(\exists x)(\exists y) \sim [(P(x) \rightarrow \sim P(y)) \rightarrow (\sim (\sim P(y) \rightarrow P(x)))] = \text{Vri} \qquad (1)$$

Then (by Rule 4), there exists a constant *a* such that:

$$(\exists y) \sim [(P(a) \rightarrow \sim P(y)) \rightarrow (\sim (\sim P(y) \rightarrow P(a)))] = \text{Vri} \qquad (2)$$

And so (following Rule 4), there exists a constant *b* such that

$$\sim [(P(a) \rightarrow \sim P(b)) \rightarrow (\sim(\sim P(b) \rightarrow P(a)))] = \text{Vri} \qquad (3)$$

But this is impossible. The two constants *a* and *b* effectively correspond, by the semantic function, to the *unique* element *u* of the universe. As such, the evaluation of $P(a)$ is exactly the same as that of $P(b)$: if [pV] is the subset of the universe corresponding to the predicate *P*, the evaluation leads us to ask whether or not the element *u* belongs to [pV] (Rule 1 of the evaluation of closed formulae).

In formula 3, we are therefore able to replace $P(b)$ by $P(a)$ without modifying the evaluation of the ensemble. The formula obtained is:

$$\sim [(P(a) \rightarrow \sim P(a)) \rightarrow (\sim(\sim P(a) \rightarrow P(a)))]$$

But this formula is *never* valid. This may be seen easily enough by 're-constructing' it. For the case where $P(a) = \text{Vri}$, for example, we may put:

Rule 2: $\sim P(a) = \text{Fax}$
Rule 3: $(\sim P(a) \rightarrow P(a)) = \text{Vri}$
Rule 2: $\sim(\sim P(a) \rightarrow P(a)) = \text{Fax}$

Let's call this result (1). Yet, if $P(a) = \text{Vri}$, we have:

Rule 2: $\sim P(a) = \text{Fax}$
Rule 3: $(P(a) \rightarrow \sim P(a)) = \text{Fax}$

Let's call this result (2). From (1) and (2), by application of Rule 3, it is possible to obtain:

$$[(P(a) \rightarrow \sim P(a)) \rightarrow (\sim (\sim P(a) \rightarrow P(a))] = \text{Vri}$$

And finally, by Rule 2:

$$\sim [(P(a) \rightarrow \sim P(a)) \rightarrow (\sim(\sim P(a) \rightarrow P(a))] = \text{Fax}$$

We will leave it to the reader to establish, by the exact same method, that if $P(a) = \text{Fax}$, one arrives at the same result. This is to say that the initial hypothesis, concerning the validity of the formula:

$$(\exists x)(\exists y) \sim[(P(x) \rightarrow \sim P(y)) \rightarrow (\sim(\sim P(y) \rightarrow P(x)))]$$

must be rejected if the universe of interpretation consists of only one element: in such a universe the formula is never valid. It therefore prescribes a *type of multiplicity* for the structure: it must possess at least two elements. It is therefore a mathematical formula, whose axiomatic markings [*marquage*] produce the theory of the structure of a set of at least two elements, requiring of it nothing else for it to be a model of the system.

We have considered the separative efficacy of an axiom, which disengages a certain type of model from the plurality of structures. The converse problem can be posed as well: a certain type of structure being provided, how do we find its syntactic signature—its adequate axiom—a formal theory taking that structure as a model? This problem is precisely that of mathematical *formalization*, where the 'providence' of models, is here the historical condition of structures: real mathematical production.

Let's return to the above example, but from the other direction: we will search for an axiom that can *only* be valid for structures whose universe consists of a single element. It is clear that for a structure of this type, the interpretation of the quantifiers is quite special: the $(\exists x)$ coincides with the $(\forall x)$, because the existence of one element of the universe belonging to a given subset entails that all the elements (there is only one) belong to it as well. Hence the idea of taking as axioms for the mathematics of the One *all* formulae of the type:

$$(\exists x)P(x) \rightarrow (\forall x)P(x)$$

where P is a predicative constant admitted by the syntax. There would therefore be as many axioms of the One as there are predicative constants.

Suppose that a structure is a model of our theory: all the axioms in question are valid. We can distinguish two cases:

1) $(\forall x)P(x)$ = Vri (in this case, after Rule 3, the axiom is effectively valid). This means that for *every* constant a, $P(a)$ = Vri. In other words (by Rule 1), *all* the elements of the universe which correspond to the individual constants belong to the subset [pV] which represents P. We will say that P is *absolute for the structure.*

2) $(\forall x)P(x)$ = Fax. In this case (by Rule 3), the axiom is not valid

unless the antecedent of the implication is likewise evaluated as Fax. This entails (by Rule 4) that there *does not* exist any constant a such that $P(a)$ = Vri, which means that *no* element u of the universe corresponding to a constant belongs to [pV]. We will say that the predicate P is *void for the structure.*

Since our list of axioms exhausts all of the system's predicates, we obtain the following result: a structure is not a model of the theory signed by axioms of the type

$$(\exists x)P(x) \rightarrow (\forall x)P(x)$$

unless *all* the predicates of the theory are either absolute or void for the structure.

It follows that the existence of different individual constants in the system's syntax has no impact on the evaluation of the formulae. Suppose, for instance, that there exist two constants a and b and a predicate P. Either P is absolute, and therefore $P(a) = P(b)$ = Vri, or P is void, and therefore $P(a) = P(b)$ = Fax. Semantically, such a theory is equivalent to the same theory deploying only a single constant.

In the same way, we could just as easily reduce the list of predicates to only two: the predicate 'absolute' and the predicate 'void'. For if P and Q are absolute, then $P(a) = Q(a)$ = Vri for the unique constant a. And if P and Q are void, then $P(a) = Q(a)$ = Fax.

Hence, the *fundamental model* of our theory, the model that evinces itself in the light of our reduced theory—reduced to a single individual constant and two predicative constants, one absolute and the other void—is the following:

— The universe, written $\{u\}$, is a set consisting of one sole element, u.
— The subsets are the empty set [*l'ensemble vide*] and the set $\{u\}$ itself.

The element u is made to correspond to the constant a; to the void predicate, the empty set; to the absolute predicate, the set $\{u\}$. That this gives us a model is a trivial matter.

We have thus demonstrated the following (weak!) theorem: a theory whose axioms are the formulae $(\exists x)P(x) \rightarrow (\forall x)P(x)$ is semantically equivalent to a theory that admits as a model a structure whose universe contains only one element. This was, by and large, the desired result. These examples suffice to show in what sense *a model is the mathematically constructible concept of the differentiating power of a logico-mathematical system.*

The double occurrence of mathematics in this statement constitutes the support of my final arguments.

9. The Category of Model and Mathematical Experimentation

The clearest lesson of our detour is that the construction of the concept of model is strictly dependent, in all of its successive stages, on the (mathematical) theory of sets. From this point of view, it is already inexact to say that the concept connects formal thought to its outside. In truth, the marks 'outside the system' can only deploy a domain of interpretation for those of the system within a *mathematical envelopment*, which preordains the former to the latter. The state of the 'productive forces' of mathematics, not mentioned as such in the interpretation, are nevertheless what condition its scientificity, and assure the unity of the plane [*plan*] on which formal syntax and 'intuitive' domains can enter into relation [*rapport*] with one another. The instruments of the correspondence are part of a mathematical theory that one must be capable of using 'naïvely'. It is effectively presupposed that a conceptual (mathematical) role is played by words or marks like set, subset, function, ∈, unions, power of a set, empty set, etc. Semantics, here, is an *intramathematical* relation between certain refined experimental apparatuses (formal systems) and certain 'cruder' mathematical products, which is to say, products accepted, taken to be demonstrated, without having been submitted to all the exigencies of inscription whose verifying constraints are governed by the apparatus.

Now, to be precise, the effectuation of semantic correspondence is *nothing other* than verification itself. It permits us to evaluate the type of scriptural rigour that can be claimed for the domain in question. Control (technical control) of the formal system permits the inscription of a *proof of deducibility* [*déductibilité*] relative to the informal demonstrations that constitute its various models.

Semantics is an experimental protocol. This does not at all mean that systems are the 'formal' of which the concrete realizations are models, but just the opposite: formal systems constitute the experimental moment, the *material* concatenation [*l'enchainement*] of proof, after the conceptual concatenation of demonstration [*aprés celui, conceptuel, des démonstrations*].

We mustn't lose sight of Lacan's fundamental theses regarding the materiality of the signifier.[1] By their light, Bachelard's celebrated definition of scientific instruments as 'materialized theories' rightly applies to these scriptural apparatuses that are formalized syntaxes: syntaxes that are in all reality *the means of mathematical production*, by the same right by which we may call vacuum tubes and particle accelerators the means of production proper to physics.

What is at stake in the technical necessity, on which we have insisted, of an effective control of syntactic procedures, and of the explicit character of the criteria for correct expressions and deduction, reflecting the function of verification-rectification that befalls formal systems, is a 'rigid' materiality that is both manipulable and open. Let us add that the increasingly evident kinship between the theory of these systems and the theory of automata, or of calculating machines, strikingly illustrates the experimental vocation of formalisms. Again, it must be well understood that materiality does not begin with machines *stricto sensu*. A formal system *is* a mathematical machine, a machine *for* mathematical production, and is placed within the process of this production.

There is, however, another essential aspect of Bachelard's definition. The scientific instrument, the means of the proof's concatenation, is itself a scientific *result*. Without theoretical optics, no microscope; without the break with the Aristotelian ideology of the 'natural plenum', no vacuum tube, etc. Let us add: without recursive arithmetic, no formal system; and without set theory, no scientific usage and no rigorous experimental protocol for these systems, and hence no system at all.

In effect, we have shown that semantic operations require a non-formalized set-theoretic material, but we could easily show that the study of the syntactic properties themselves requires fragments of the theory of integers, and notably—we have mentioned it in passing—a constant utilization of recursive reasoning over the length of inscriptions. These are—among others—the regions of mathematical science *incorporated* into the material apparatuses where this science is put to the test. These

1. Jacques Lacan, *Écrits*, especially, 'L'Instance de la lettre dans l'inconscient', [Paris, Éditions du Seuil, 1966,] pp. 493-528, and 'La séminaire sur "La Lettre volée"', pp. 11-60. [See *Écrits*, trans. Bruce Fink, New York, W. W. Norton, 2006.]

incorporations attest to the fact that the means of mathematical production are themselves mathematically produced, that they themselves are rooted in the 'double occurrence' of mathematics in our definition of the concept of model. Far from indicating an outside of formal thought, the theory of models governs a dimension of the sciences' *practical immanence*—a process, not only of the production of knowledge, but of the reproduction of the conditions of production.

In the unity of this process, the distinction between syntax and semantics has the fragility of the distinction between the *existence* and the *use* of an experimental apparatus. This distinction has value only so long as one specifies the apparatus' incorporation of scientific regions which are not directly at stake in the proof in which the apparatus figures. For the same reason, a decisive advance in our knowledge of viruses must await the optical perfections of a microscope.

Likewise, the relevant distinction between semantics and syntax refers to the choice of which part of mathematics is to figure *in the metalanguage*.

We will now call 'metalanguage' all that is required of the current (non-formalized) language, consisting here of 'intuitive' mathematics, for it to be possible to rationally explain and practice the syntactical and semantical operations.

From this point of view, what must essentially be said is that the discipline of syntax is arithmetical, and that of semantics, set-theoretic. Remember: the theory of apparatuses of inscription, conceived as mathematical *objects*, borrows the essentials of its concepts from recursive arithmetic—or the arithmetic of transfinite ordinals. These arithmetics effectively permit the ordering, and the inductive numbering, of the experimental *assemblage*, as when they evaluate strength, complexity, etc., by reasoning over the structure of the inscriptions that the system authorizes or rejects. In return, the theory of the apparatus' *uses*, understood as experimental operations, seeks to *classify* regions of the material-mathematics of the mathematics treated in the apparatus: this aim is likewise that of the concept of structure, itself produced in the most general, the most enveloping, of theories at our disposal: set theory (or now: category theory).[2]

2. The mathematical concept of a category is a generalizing reconfiguration [*refonte*] of the concept of a species of structure. One may refer to: Georges Poitou (Cours polycopié), *Introduction à la théorie des catégories*, [Paris, Offilib, 1967,] chapters 1 and 2. Concerning the possibility of developing all of known mathematics in the language of categories, F. William Lawvere, 'The Category of Categories as a Foundation for Mathematics', in *Proceedings of the Conference on*

This side of things has been perceived in part by Kreisel and Krivine in their *Elements of Mathematical Logic* (1967), whose subtitle is precisely: *Model Theory*. Recalling the (ideological) terminology specific to the 'foundations of mathematics', they distinguish two perspectives:

— the 'set-theoretic, semantic foundations', whose 'basic notions are: set, the membership relation (between sets) and the 'logical' operations (on sets) of union, complementation and projection'.[3]

— the 'combinatorial foundations', of which 'the basic notions are: *word*, i.e. a finite series of symbols of a finite alphabet, *combinatorial function* (whose arguments and values are words), and *combinatorial proof* of identities (between differently defined combinatorial functions ...)'.[4]

In both cases, the authors emphasize the dominant mathematical reference from which each perspective originates: Semantics is realist; it 'accept[s] set theoretic terminology as meaningful, and not only as a 'façon de parler'.[5] Combinatorics rests upon (arithmetical) notions that are 'quite familiar [...] because [they are] involved in all elementary mathematics'.[6]

But for want of having done with the unilateral ideology of 'foundations', Kreisel and Krivine fail to seize this difference as a moment of a unique experimental process, where combinatorics is only the experimental assemblage for a scriptural verification, the practical forms of which are governed by semantics. They are therefore reduced to giving their *opinion* on the respective merits of each approach, the separation of which is a mere impotence.

Nevertheless, they clearly designate the sole support for thinking the difference/unity of the model and the formal, of semantics and syntax: it is the intra-mathematical relation [*rapport*] between arithmetical and set-theoretic 'raw materials'.

Insofar as the concept of model articulates this difference, we should expect that the theoretical results would concern something *adhering* to

---

Categorial Algebra. [Today, readers may also consult the remarkable text by F. W. Lawvere and Robert Rosebrugh, *Sets for Mathematics*, New York, Cambridge University Press, 2003. For Badiou's own use of category theory in his recent work, see *Court traite d'ontologie transitoire* and *Logiques des mondes: l'être et l'événement 2*.]

3. [G. Kreisel and J. L. Krivine, *Elements of Mathematical Logic (Model Theory)*, Amsterdam, North-Holland Publishing Company, 1967, p.166.]

4. [Kreisel and Krivine, p.195.]

5. [Kreisel and Krivine, p. 166.]

6. [Kreisel and Krivine, p. 195.]

mathematical practice, and not authorize any exportation. This is not only because these results concern mathematical experiments, but because the rule of use of the word 'model', and the principles engaged in the demonstrations in which this word appears, refer to the conceptual systems of mathematics.

This is in fact the case: the fundamental theorem of completeness, for a system of the type that has provided me with an example, states that such a system is consistent if and only if it possesses a model. (cf. the appendix). This theorem ties a syntactical concept (consistency) to a semantical concept (model). Within the project of the epistemology of models, it stands as a crucial point in the *juncture* between the 'formal' and the 'concrete'. But its demonstration requires one to be able to well-order *all* the correct formulae of the system, which, in general, requires a very strong set-theoretical proposition: the axiom of choice. The completeness theorem has meaning only in the workspace of mathematics. In fact, it is a theorem of set theory, or rather of *a* set theory, for we have known since the work of Cohen that the axiom of choice is independent of the other axioms, so that it is possible to construct a set theory where the axiom is explicitly negated. This is to say that every exportation outside of the domain proper to mathematical experimentation is illegitimate, so long as one presumes to guard the rigour of the properties of the concept, and not degrade them into variants of an ideological notion.

We have thus established that the philosophical *category* of model, such as it functions in the discourse of logical positivism, is doubly inadequate.

It is inadequate, first of all, in that it pretends to think science in general according to a difference (syntax/semantics) that is itself nothing more than an ideological *relapse* [*rechute*] of a regional intra-mathematical difference (between recursive arithmetic and set theory).

It is above all inadequate insofar as it pretends to outfit empiricist ideology in *words* which designate moments of a mathematical process. In its discourse, 'formal languages' and 'empirical facts' effectively confront one another as two heterogeneous regions. That the latter are eventually 'modelled' in the former permits the confrontation to be 'thought' as a relation [*rapport*]. But in mathematics, to be precise, the formal apparatus is that by which a mathematical region, in taking its place [*advenant*] as a model, finds itself *transformed*, tested, and experimented upon, as concerns the state of its rigour or generality. It is inconceivable that there should be a similar transformation of something other than that which, being always *already* mathematical, is semantically assignable as capable of being articulated together with a syntactic apparatus. It is because it is itself a materialized theory, a mathematical result, that the formal apparatus can enter into the process of the production of mathematical knowledge; and in this process, the concept of model does not designate an outside to be formalized, but a mathematical material to be tested.

The discourse of Carnap, like that of Lévi-Strauss, is a variant of bourgeois epistemology. It exhibits a combination of empiricist notions pertaining to the 'problem of knowledge' and scientific concepts borrowed from mathematical logic, a combination that defines a philosophical *category* of model in which ideology is dominant and science servile.

10. The Category of Model and the Historical Time of Mathematical Production

Is this to say that no epistemological usage of the word 'model' is admissible? Assuredly not, if it is simply directed to the *historicity* of mathematics, in the form of their experimental dialectic. The category of model then serves to think the (very particular) time of this history.

Allow me to clearly specify the scope [*portée*] of this development: I am obviously not pretending to draw a doctrine of the history of mathematics out of the concept of model. Quite to the contrary, such a doctrine cannot appropriate the category of model save insofar as it *already* implicitly commands both a polemic against the notional (ideological) uses of the term, and an analysis [*lecture*] of the (scientific) concept.

I am saying only this: if one assumes, within the framework of dialectical materialism, a doctrine of the *historical production* of scientific knowledge, one is led to recognize, in the concept of model, an *epistemological index*, on the basis of which the experimental dialectic of mathematical production is deciphered, and is extracted from its idealist ordinance in 'pure', 'formal', or 'a priori' knowledge, etc.

In other words, once clarified by dialectical materialism, the rigorous examination of the scientific concept of model permits us to trace a line of demarcation between two categorial (philosophical) *uses* of that concept: one is positivist, and enslaves it to the (ideological) notion of science as representation of the real; the other is materialist, and, according to the theory of the history of sciences (a specific region of historical materialism) indirectly readies its effective integration into proletarian ideology.

Finally, the uses of the word 'model' may find themselves distributed in a table such as that on page 49, at the centre of which one finds the

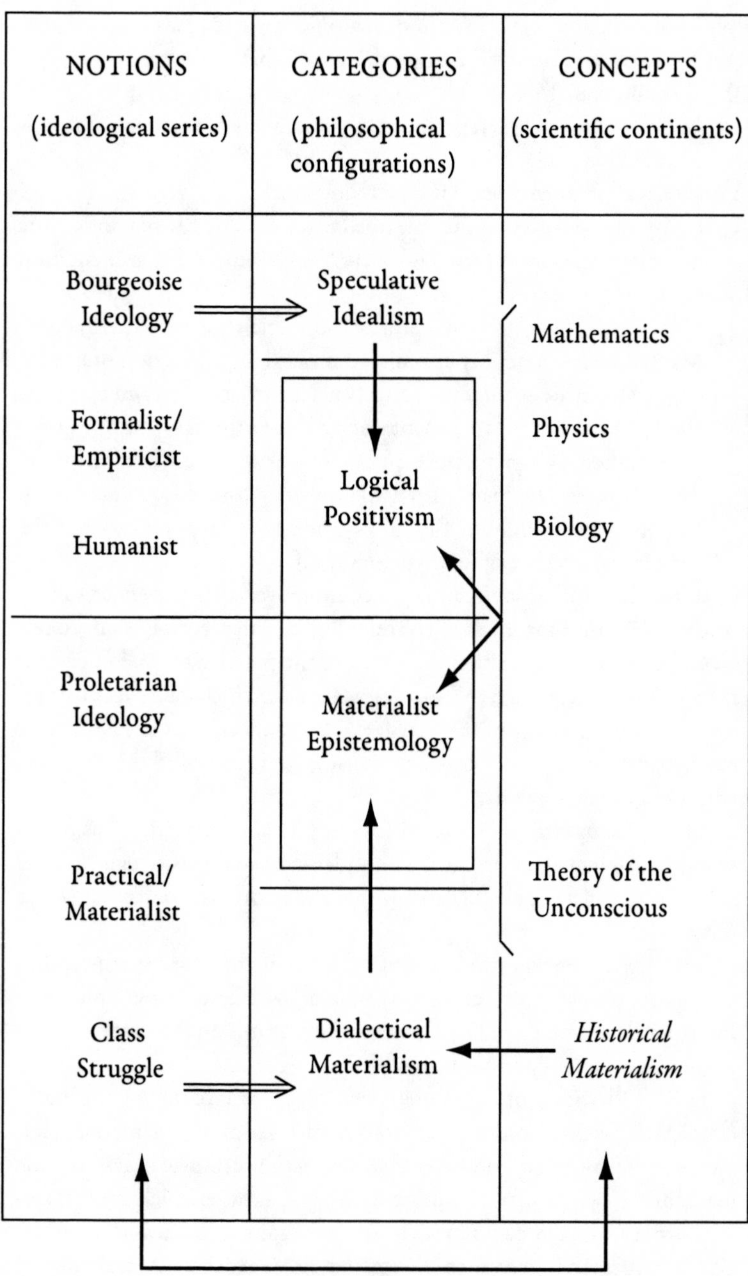
NOTIONS
(ideological series)
CATEGORIES
(philosophical configurations)
CONCEPTS
(scientific continents)
Bourgeoise Ideology
Speculative Idealism
Mathematics
Formalist/ Empiricist
Physics
Logical Positivism
Humanist
Biology
Proletarian Ideology
Materialist Epistemology
Practical/ Materialist
Theory of the Unconscious
Class Struggle
Dialectical Materialism
Historical Materialism
Proletarian Politics

epistemological struggle which effectively concerns the entirety of the 'philosophy course' of which our development is but a part.

Of the acceptations of the word 'model' that are at stake here, we should enumerate four:

1) *Notion*: knowledge is representation by models of the real-empirical-given.
2) *Concept*: (mathematical): model theory.
3) *Category* 1 (positivist): the empirical real furnishes the semantics (the models) of the syntax proposed by the 'pure' sciences. Experimentation is an evaluation/realization.
4) *Category* 2 (dialectical materialist): all sciences are experimental. Mathematics is a doubly articulated process of production of knowledge. 'Model' designates the conceptual articulation, *insofar* as it is related to a particular experimental apparatus: a formal system. 'Formal system', then, designates the experimental articulation, or inscription. There is an envelopment of articulation-2 by articulation-1: the intellection [*l'intelligence*] of formal assemblages is deployed within the conceptual practice of mathematics itself.

In reading this table it can be seen, moreover, that the effect sought by the (dialectical materialist) epistemological intervention is not to put an end to that which defines philosophy: the practice of an 'impossible' rapport between science and ideology. That which characterizes this intervention is, in fact, its reflective [*réfléchi*] rapport with *one* altogether particular science: historical materialism; and, conjointly, its rapport with proletarian ideology.

In the last instance, the line of philosophical demarcation has for its practical referent the class struggle in ideology, and this struggle has its *stakes* in the class-appropriation [*l'appropriation-de-classe*] of scientific practice.

This general background, which determines the Marxist conception of philosophy, can only be violently schematized here. For the moment, I will perilously confine myself to a few indications on the correct [*juste*] usage of the category of model in epistemology.

First of all, the theory of models permits, as we have shown, the mathematical differentiation of logic from mathematics. It organizes an employment of formal apparatuses that allows the formulae specifying the mathematicity of a structure to be located—as those which *prevent* certain structures from being models for the system. Now this differentiation inscribes itself in a venerable epistemological debate (what is logical and 'universal', and what is mathematical and regional?), which it diversifies and rationalizes.

However, the principal use of models is bound up with the production of proofs of *relative consistency* and *independence*.

Let *T* be a formalized theory defined by its axioms, and let *A* be a well-formed expression in the formal language adopted. Call (T + *A*) the theory obtained by the adjunction of *A* to the axioms of T. One will say that the formula *A* is *consistent* with T if, presuming that T is consistent, (T + *A*) is as well. How can such results be established, whose appearance is purely syntactic?

The fundamental theorem of completeness guarantees us that a theory is consistent if and only if it admits of a model. The hypothesis concerning the consistency of T requires that we consider this theory as the experimental inscription of a structure. In 'building' [*travaillant*] this structure—in developing the supposed consistency of T—we seek to produce a model of (T + *A*), that is, a structure which would be a model of T and in which, moreover, *A* is valid. The consistency of (T + *A*) is thus guaranteed.

These are the means by which Gödel demonstrated, in 1939, the consistency of the axiom of choice and the continuum hypothesis relative to set theory with neither the axiom of choice nor the continuum hypothesis.[1]

But this demonstration's point of interest, its *epistemological weight*, lies in the fact that the axiom of choice had been disputed, even rejected, by a number of mathematicians and logicians, who nevertheless admitted the rest of the theory. Their suspicion was bound up with a certain vision of mathematics, which privileged 'effective' operations and the foundational function of the integers. It thereby depended on a (philosophical) category: a category separating what is mathematical, or rational, from what is not.

The Gödelian experimentation—in which the formal apparatus, here being the axiomatization of set theory, plays a decisive role—*intervenes* in an epistemological conjuncture by means of science. It proves that the axiom of choice is not, from the point of view of consistency, any more 'risky' than the rest of set theory. It dissipates suspicion, and warrants use. This being done, it transforms, not the theory, but its *status* in the historical process of the production of knowledge: the once disquieting

1. [For a discussion of these proofs in the context of Badiou's later philosophical thought, see Meditation Twenty-Nine ('The Folding of Being and the Sovereignty of Language') of *Being and Event*.]

question of knowing whether or not the demonstration of such and such a proposition is independent of the 'suspect' axiom, essentially ceased to be of interest.

Without a doubt such an intervention, in virtue of the very minutiae of the experimental assemblages that it demands, always comes after the fact [*aprés-coup*]. *Practice* had already settled in favour of the axiom of choice. But the intervention, to be precise, altered this 'choice' by the test to which it submitted it. In a sense, it confirmed and established [*trouve établi*] that it was less of a 'choice' than a necessity internal to the mathematical process. As in physics, the *belatedess* [*retard*] *of the (experimental) proof* operates retroactively upon *mathematical anticipation*.

We will now turn to a classic example. Let EG be the (supposedly formalized) theory of Euclidean spatial geometry. We suppose it to be consistent, and so, after the completeness theorem, it admits of a model. To simplify things, we will consider this model to be Euclidean space, such as we have an academic 'intuition' of it (but, here, these are names for complex structures expressible in the language of set theory).

Now take the theory obtained by replacing, in Euclidean *plane* geometry (a sub-theory of EG), Euclid's celebrated postulate: 'through a point exterior to a straight line, there passes one and only one straight line parallel to the first', by the axiom (which implies the negation of the former): 'through a point exterior to a straight line, there pass *no* straight lines parallel to the first'. We will call this new theory RPG (Riemannian plane geometry).

We are going *to interpret RPG in a structure constructible on the basis of the model of EG*. In this model, whose universe is a Euclidean space, let there be a (Euclidean) *sphere*. It will be the universe of our sub-structure.

— To the constants of RPG which mark the points (of a plane), the points of the sphere will be made to correspond. But we will hold *two diametrically opposed points to be identical*: the 'elements' of our structure are therefore *pairs* of points.
— To the constants of RPG which mark the straight lines (of a plane), the great circles of the sphere (circles whose plane passes through the sphere's centre) will be made to correspond.
— The relation between straight lines: 'having one point in common' is interpreted without any changes.

We may easily verify that this structure is a model for the 'normal' axioms of RPG. For example, the axiom (common to RPG and EG), 'through two points there passes one and only one straight line', is interpreted, 'through two different, non-diametrically opposed points of the sphere, there passes one and only one great circle', which is true in every model *of EG* (it is a theorem of EG, or rather of its interpretation).

One may also ascertain that the axiom that characterizes RPG (the non-existence of parallels) is valid for this structure, since two great circles *always* intersect.

If, therefore, EG admits of our model, we can then construct a model for RPG. From this apparatus, it follows that the consistency of RPG is guaranteed by that of EG.

It thus follows that Euclid's famous postulate is independent from the other axioms of EG. If one could in fact deduce it from these axioms, then every model of (EG—*A*)—the formal inscription of Euclidean geometry minus the postulate—would *also* be a model of EG, since deduction conserves validity. But our model of RPG *is* a model of (EG—*A*), because all the axioms other than the Euclidean postulate are conserved in RPG, and are all consequently valid for the sphere-structure. Yet this structure is certainly not a model of EG, since the *negation* of the postulate is valid within it. It follows that we cannot expect to deduce this postulate (invalid for a certain structure) from the other axioms (valid for this same structure).

It is thus that in producing a Euclidean model for Riemannian geometry, Poincaré retrospectively *bolstered* [*étayait*] the advance of the 'new' geometries over the concepts of classical geometry, whose venerable practice seemed to exempt it from any suspicion of inconsistency.

And likewise, it is thus that the model, by the independence proof that it administered, retrospectively transformed the status of those vain efforts deployed, over the course of centuries, to demonstrate the Euclidean postulate: their defeat was necessary, and not a matter of circumstance. It was a matter of impossibility, and not impotence. In the same stroke, the model put an end to the practice that it judged.

This leads us to the true import of the category of model.

Take a given mathematical configuration, inscribed in the history of the science. To make this configuration appear as a model of a formal system—to process it through this mechanism—has the principal effect of *placing* it in its particularity, of exporting it, outside of the immediate illusions of its singular production, into a more general mathematical space, that of the system's models, in the plural: the experimental apparatus is a crossroad of practices.

These operations of placement can be historically decisive: at the beginning of the nineteenth century, little was known of groups beyond their calculus of substitutions; the progressive disengagement of the axioms of group structure resulted from scriptural manipulations which led to the appearance of the 'substitution groups' as models among others. The impetus [*élan*] that this generalization would give to algebra throughout the length of the century is well known.

However, as a mathematician once pointed out to me, the true problem posed by this impetus is that the generalization in which it resulted is only apparent: it is well known, in fact, that *every* group is isomorphic to a substitution group. It is thus that formalism is the retrospective test of the concept. It commands the time of the proof, not that of the demonstrative entanglement. The placement that it put to work under the concept of model's jurisdiction *readjusts the concepts treated with respect to their own implicit powers*. Identical and displaced, the concept of a substitution group has traversed the experimentation whose specific assemblage was the formal theory of groups in general [*des groups quelconques*]. Its importance thus comes to be *verified*, already marked in its practical predominance at the beginning of the nineteenth century, and the type of generality to which it can lay claim is *rectified*.

This use of the word 'model', to my mind, delivers a fertile epistemological category. I propose to call *model* the ordinance [*statut*] that, in the historical process of a science, retrospectively assigns to the science's previous practical instances their experimental transformation by a definite formal apparatus.

Reciprocally, the conceptual historicity, which is to say the 'productive' value, of formalism comes to it both from its theoretical subordination [*dependence*] as an instrument, and from what it has as models: from what it doubly incorporates into the conditions of production and reproduction of knowledge. Such is the *practical* guarantee of formal assemblages.

The category of model thus designates the retroactive causality of formalism on its own scientific history, the history conjoining object and use. And the history of formalism will be the anticipatory intelligibility of that which it retrospectively constitutes as its model.

The problem is not, and cannot be, that of the representational relations between the model and the concrete, or between the formal and the models. The problem is that of *the history of formalization*. 'Model' designates the network [*réseau*] traversed by the retroactions and anticipations that weave this history: whether it be designated, in anticipation, as break [*coupure*], or in retrospect, as remaking [*refonte*].[2]

2. François Regnault, *Cours de philosophie pour scientifiques*, Installment III.
[In the foreword to Pêcheux and Fichant's *On the History of the Sciences* (*Sur l'histoire des sciences*, Paris, F. Maspero, 1969), the editors of *Théorie* include the following note:

> As a result of circumstances independent of our wishes, it is impossible for us to publish the course which F. Regnault gave on the 26th of February, 1968, bearing the title 'What is an Epistemological Break?' [« *Quést-ce qu'une coupure epistemologique ?* »], within the framework of the 'Philosophy Course for Scientists'.

However, the following explanation is provided in a footnote to 'Infinitesimal Subversion', where Badiou writes:

> We have taken from F. Regnault the concept of remaking, by which he designates those great modifications where, returning to the unthought of its preceding epoch, a science globally transforms its conceptual system—e.g. relativistic mechanics after classical mechanics (p.120).]

Appendix

§1: The goal

My intention here is to provide a few pieces of information on the completeness theorem—notably those which concern purely logical theories constructed in the language of my basic example. I refer the reader again to the supplement at the end of the text.

These remarks, besides exercising the reader in the 'back and forth' movements characteristic of semantic methods and reasonings of the syntactic sort (recursion on the length of inscriptions), have the merit of legitimating the example under consideration. One form of the theorem is, in fact, the following: every theorem and axiom of the system is valid for every structure; reciprocally, every formula valid for every structure is an axiom or a theorem of the system.

This system thus permits us to deduce all purely logical formulae expressible by means of the stock of marks available, and by them alone. Any structure whatsoever is a model for this system. This semantico-syntactic equivalence assures us that our apparatus is a complete formal logic (at this level, which admits predicates with only one argument).

The task at hand is less to give a complete demonstration of the theorem, or an inventory of the most efficacious methods, than to run through certain common procedures, according to deliberately slowed and accelerated rhythms. In principle, a little attention should suffice; nothing else is required. The reader will occasionally be left to finish the proofs as an exercise.

§ 2: Description of the apparatus PS

We will call this system PS (predicative system). The stock of marks and the rules of formation will be those of our example (cf. the supplement at the end). The axiom schemata will be the following (unless otherwise specified, A and B are any well-formed expressions whatsoever):

Ax 1: $A \rightarrow (B \rightarrow A)$
Ax 2: $[A \rightarrow (C \rightarrow D)] \rightarrow [(A \rightarrow C) \rightarrow (A \rightarrow D)]$
Ax 3: $A \rightarrow (\sim A \rightarrow B)$
Ax 4: $(A \rightarrow \sim A) \rightarrow \sim A$
Ax 5: $\sim\sim A \rightarrow A$
Ax 6: $(\forall x)A \rightarrow A(f/x)$, where x is free in A, and where f is either a constant or a variable unbound in the part of A in which x is free.
Ax 7: $(\forall x)(A \rightarrow B) \rightarrow [A \rightarrow (\forall x)B]$, if x is not free in A.

We are not asking the question of whether or not these axioms are independent of one another. In fact, they are not: axioms 3 and 4 can be deduced from axioms 1, 2 and 5. But our choice of axioms simplifies the demonstrations.

One may be surprised that none of these axioms mention the existential quantifier. This is because that quantifier is definable on the basis of the universal and negation. The assertion, 'there exists an x possessing the property P' is (semantically) equivalent to the assertion: 'it is false that every x is marked by not-P'. We will therefore consider $(\exists x)A$ to be merely an inscription abbreviating $\sim(\forall x)\sim A$. In what follows, we will consider every quantified expression to be exclusively composed of universal quantifications.

The rules of deduction for PS are those which we have already mentioned: the rule of separation and the rule of generalization. The description (the assemblage) of the system is thus complete.

§ 3: Every theorem of PS is purely logical

Our intention is to establish that every formula that can be deduced in PS is valid for every structure. For this purpose, it suffices to verify that this is true for the axioms, and that the rules of deduction conserve validity. We'll call the property of being valid in every structure 'L-validity' (logical validity).

So far as the axioms are concerned, we'll leave some of the work to the readers. I have already shown in the text that the schema $A \rightarrow (B \rightarrow A)$ is always valid. The method is the same for axioms 2, 3, 4 and 5 (the repeated employment of semantic rules 2 and 3). For axiom 6, one may see that it is certainly L-valid after rule 5.

Let's put the case of axiom 7 to the test. If it is not L-valid, there exists a structure in which a closed instance of this axiom takes the value Fax.

Assuming that A does not contain the free variable x, this is written:

$$(\forall x)(A' \rightarrow B') \rightarrow [A' \rightarrow (\forall x)B'] = \text{Fax} \qquad (1)$$

where A' is a closed instance of A and B' is a formula whose only free variable is x.

If (1) is verified, Rule 3 necessitates:

$$(\forall x)(A' \rightarrow B') = \text{Vri}$$

and so for every constant a (Rule 5):

$$(A' \rightarrow B'(a/x)) = \text{Vri} \qquad (2)$$

At the same time, Rule 3 necessitates that:

$$\begin{array}{lll} & (A' \rightarrow (\forall x)B') & = \text{Fax} \\ \text{since (by Rule 3)} & A' & = \text{Vri} \\ \text{and (Rule 3 again)} & (\forall x)B' & = \text{Fax} \end{array} \qquad (3)$$

which (by Rule 5) means that for at least one constant a:

$$B'(a/x) = \text{Fax} \qquad (4)$$

If equations (3) and (4) are satisfied, (2) cannot be. The hypothesis must be rejected, and our axiom is L-valid. As for the rules of deduction, we have already shown that they conserve validity (cf. CM 33-4).

We have therefore ascertained that, starting from axioms valid for every structure, we exclusively deduce formulae that are valid for every structure. Our system PS contains no mathematical deductions: it does not experiment on structural differences. It is a logical machine.

It remains to establish that this machine exhausts the logical domain expressible by its resources of inscription. Or, in other words, that every L-valid formula is deducible in PS. This point is a much subtler one, and requires a few detours.

§ 4: Deduction theorem

In informal mathematical practice, in order to establish a theorem, one frequently has need of a condition that is *supplementary* with respect to the structural generality in which one is working. This is the famous scholarly use of 'hypotheses': if I suppose the statement A, then I can demonstrate statement B.

Apparently, this translates into our logical language as the formula:

$A \rightarrow B$. But only apparently. The supposition, in fact, has no place in a formal system. $(A \rightarrow B)$, in all rigour, means: if I *deduce* A and $(A \rightarrow B)$, then I can deduce B. This has the same sense as the rule of separation. But if, in PS, I cannot deduce A, then the deduction of $(A \rightarrow B)$ says nothing about the deducibility of B. How, then, are we to translate the idea that the hypothesis A permits us to conclude something about B?

We are going to show that our system is able to inscribe this problem.

At bottom, supposing that A is true comes down to adding it to the axioms. Let (PS + A) be the system obtained by adjoining the 'hypothesis' A to the axioms of PS. For the purpose of simplification, we will consider only closed formulae A. We then have the following result, which in fact drives the deductive efficacy of the apparatus:

Deduction Theorem: If the formula B is deducible in the system (PS + A), the formula $(A \rightarrow B)$ is deducible in the system PS.

Consider any deduction whatsoever in the system (PS + A). It is a finite sequence of formulae, which we enumerate (in deductive order) in the following fashion: $B_1, B_2, B_3, ..., B_n$. We will reason by recursion in order to establish that $(A \rightarrow B_n)$ is a theorem of PS (without the axiom A).

First of all let's examine the case of B_1, the first formula of the deduction in (PS + A). Every deduction begins with an axiom: B_1 is therefore either an axiom of PS, or the supplementary axiom A.

— If B_1 is an axiom of PS, we have the following deduction *in* PS:

$$
\begin{array}{ll}
B_1 & \text{(axiom by hypothesis)} \\
B_1 \rightarrow (A \rightarrow B_1) & \text{(axiom 1)} \\
\quad A \rightarrow B_1 & \text{(separation)}
\end{array}
$$

— If B_1 is the supplementary axiom A, then we will leave the reader the trouble of verifying that the following sequence is a deduction in PS:

$$
\begin{array}{l}
A \rightarrow [(C \rightarrow A) \rightarrow A] \\
[A \rightarrow [(C \rightarrow A) \rightarrow A]] \rightarrow \; [[A \rightarrow (C \rightarrow A)] \rightarrow (A \rightarrow A)] \quad \text{(Ax. 2)} \\
\qquad [A \rightarrow (C \rightarrow A)] \rightarrow (A \rightarrow A) \\
\qquad A \rightarrow (C \rightarrow A) \\
\qquad\qquad A \rightarrow A \\
\qquad\qquad A \rightarrow B_1
\end{array}
$$

Hence $(A \rightarrow B_1)$ is always deducible in PS.

We will now formulate *the hypothesis of recursion*. We assume that for every formula B_i preceding B_n in a deduction in (PS + A), the formula $(A \rightarrow B_i)$ is deducible in PS. We will then show that $(A \rightarrow B_n)$ is likewise deducible in PS. In (PS + A), one may produce B_n in three fashions:

a) B_n is an axiom of (PS + A), and therefore either an axiom of PS, or else the axiom A. In this case, the reasoning applied above to B_i shows that $(A \to B_n)$ is deducible in PS.

b) B_n is produced by the rule of separation. In this case, there exist formulae $(B_i \to B_n)$ and B_i preceding B_n in the deduction (in the system (PS + A)). We therefore have the following deduction in PS:

$A \to (B_i \to B_n)$ (after the hypothesis of recursion.)
$[A \to (B_i \to B_n)] \to [(A \to B_i) \to (A \to B_n)]$ (axiom 2)
$(A \to B_i) \to (A \to B_n)$ (separation)
$(A \to B_i)$ (hypothesis of recursion)
$(A \to B_n)$ (separation)

c) B_n is produced by the rule of generalization. There then exists some B_i preceding B_n in the deduction, with B_n being written: $(\forall x)B_i$. We then have the following deduction in PS:

$A \to B_i$ (by the hypothesis of recursion)
$(\forall x)(A \to B_i)$ (rule of generalization)
$(\forall x)(A \to B_i) \to [A \to (\forall x)B_i]$ (axiom 7).

We know that A is applicable, for, since A is a closed formula, x cannot be free in it.

$A \to (\forall x)B_i$ (separation)
$A \to B_n$ (inscription of B_n)

Let's put together our results: given a deduction in the system (PS + A), the first formula of this deduction, B_1, is such that $(A \to B_1)$ is a theorem of PS.

And if the formulae which precede B_n have this property, then B_n has it as well.

Every deduction being finite, a theorem of (PS + A) always has a (finite) rank n in deduction. The metatheoretical, informal usage of the schema of recursive reasoning authorizes us to conclude:

If a formula B is deducible in the system (PS + A) where A is a closed formula, then $(A \to B)$ is deducible in PS.

§ 5: The relative consistency of certain extensions of PS

Suppose that the closed formula $\sim A$ *is not* deducible in PS. We add the formula A to the axioms, and thereby obtain a new theory, (PS + A). We are going to show that this theory is *consistent*.

Recall that a theory is consistent if there exists at least one formula A that cannot be deduced from the theory. So, if (PS + A) were inconsistent, we could then deduce from it any formula whatsoever, and, in particular, the formula $\sim A$.

But, if $\sim A$ is deducible in (PS + A), the deduction theorem guarantees that $(A \rightarrow \sim A)$ is deducible in PS. But

$$(A \rightarrow \sim A) \rightarrow \sim A$$

is an axiom of PS (axiom 4). By separation, $\sim A$ would therefore be deducible in PS. As we have just assumed that it is not, the hypothesis that (PS + A) is inconsistent must be rejected.

If the negation of a closed formula A of PS is not a theorem of PS, the system (PS + A) is consistent.

§ 6: The scope of the completeness theorem

If we are to arrive at a demonstration of the completeness theorem—the theorem being that every consistent theory admits of a model—we will be assured that our system PS is indeed a complete deductive logic, or, in other words, that every closed formula valid in every structure is a theorem of the system.

Let A be a closed, L-valid formula. $\sim\sim A$ is another (semantic Rule 2). If A is not deducible in PS, then neither is $\sim\sim A$. In fact, if $\sim\sim A$ is deducible,

$$\frac{\begin{array}{ll}\sim\sim A \rightarrow A & \text{(axiom 5)} \\ \sim\sim A & \end{array}}{A \quad \text{(separation)}}$$

is a deduction in PS, and A is a theorem, contrary to the hypothesis. But if $\sim\sim A$ is not deducible in PS, then the theory (PS + $\sim A$) is consistent (theorem of the previous section). It therefore admits of a model if the completeness theorem is true. In this model, $\sim A$, an axiom of the theory

$$(\text{PS} + \sim A)$$

is evidently valid (by the definition of model). Since A is assumed to be L-valid, it is, in particular, valid for the structure that is this model. But the two formulae A and $\sim A$ cannot be simultaneously valid in the same structure: our initial hypothesis must be rejected; if A is L-valid, it is certainly a theorem of PS.

Hence, under the condition of the completeness theorem, every purely logical formula of PS is deducible in PS.

In passing, let's note that this result, like the one before it, and like the deduction theorem, holds for every theory containing the axioms of PS. It therefore holds, in particular, for mathematical theories obtained by adjoining axioms which are not purely logical to PS. That is to say, it holds for experimental mathematical apparatuses whose subjacent logic is articulated by PS.

§ 7: The Lindenbaum lemma

An interesting syntactic question concerning the experimental powers of a formalized theory is that of *saturation*: does the apparatus permit one to effectively classify *all* closed formulae as either demonstrable or refutable (a formula is refutable if its negation is demonstrable)? If such is the case, then the theory is said to be *saturated*. Given a closed formula A, either A is a theorem, or else not-A is a theorem.

Let's note right away that, for a purely logical system such as ours, (syntactic) *saturation* entails (semantic) *completeness*. In fact, if a formula A is L-valid and is not a theorem, then $\sim A$ is a theorem (saturation). But then $\sim A$ is L-valid, because our system is purely logical. As it is impossible for A and $\sim A$ to both be L-valid, our initial hypothesis is untenable: it is necessary to admit every L-valid formula as a theorem. The system is therefore complete for purely logical formulae.

As a general rule, the question of knowing whether a determinate mathematical theory is saturated is not a simple one. One famous result in this direction is due to Gödel, and concerns a formal apparatus of arithmetic: this apparatus *is not* saturated. Gödel explicitly constructed an *undecidable* closed formula (neither it nor its negation are deducible, at least so long as the system is consistent). This formula, however, can indeed be evaluated in the apparatus' 'normal' model: the natural integers equipped with their usual operations. In this model, the negation of the undecidable formula is valid. This is to say that the formal system of arithmetic is semantically incomplete with respect to its normal model.

We will now establish the following general result: *Every consistent theory admits of a saturated extension* (Lindenbaum).

A theory T will simply be a system admitting all the axioms of PS, with the possible addition of other axioms.

An extension of a theory T will be another theory T', such that all the theorems of T are also theorems of T'. T' is expressed in the same language as T, and therefore has the same well-formed expressions. The Lindenbaum lemma plays a decisive role in model theory. In the elementary version given here, it essentially depends upon the finitude of the sequences of marks (formulae), and on the idea that the stock of marks at our disposal is denumerable.

We suppose, in effect, that we can rank and enumerate *all* the closed formulae of T. Let $F_1, F_2, ..., F_n, ...$ be this ordering. Given that each formula F_n is a finite sequence of marks, and the marks themselves are capable of being numbered (they are denumerable), this supposition is justified.

We then examine the formulae in order, one after the other, to recursively define a *sequence of theories*:

— The theory T_0 is the theory T itself.

— If $\sim F_1$ is deducible in T_0, then T_1 is the theory T_0,
if $\sim F_1$ is not deducible in T_0, T_1 is the theory $(T_0 + F_1)$.

— If $\sim F_2$ is deducible in T_1, T_2 is the theory T_1,
if $\sim F_2$ is not deducible in T_1, T_2 is the theory $(T_1 + F_2)$.

....

— If $\sim F_n{+}1$ is deducible in T_n, then $T_n{+}1$ is the theory T_n,
if $\sim F_n{+}1$ is not deducible in T_n, then $T_n{+}1$ is the theory $(T_n + F_n{+}1)$.

....

The reader may make use of the result in § 5 to show that if the theory T_n is consistent, then so is the theory $T_n{+}1$. If, therefore, T_0, it i.e. T, is consistent, the recursion permits us to conclude that every theory T_n in the sequence is as well.

Consider the theory T' obtained by taking all the axioms of all the theories $T_0, T_1, ..., T_n$ ... This theory is also consistent, if T is consistent, as can be verified. Moreover, it contains (among other things) all the axioms of T_0, and therefore all of its theorems. It is indeed an extension of T. It remains for us to establish its saturation.

Let F_n be any formula whatsoever, given along with its rank n in the enumeration. Either $\sim F_n$ is deduced from the axioms of the theory T_{n-1}, or it is not. If it is, then it is thereby a theorem of T', which contains all of the axioms. If it is not, then the rule of construction for the sequence of theories tells us that T_n is $(T_{n-1} + F_n)$. F_n is therefore an axiom of T_n, and therefore of T'. Consequently, whatever F_n may be, either $\sim F_n$ or F_n is deducible in T', which is a saturated theory.

Note that this theorem is properly semantic insofar as it is not effective. It may well be impossible to decide in advance, by an invariable mechanical procedure or scriptural assemblage, whether or not, at a stage of rank *n*, the formula $\sim F_n+1$ is deducible in the theory T_n. If such a thing is possible, then the theory T_n is *decidable*: this is a very strong property for a formal apparatus, but unfortunately quite rare. PS, for instance, is decidable, but, for a theory admitting of binary relations—expressions of the type R(*x*, *y*)—with the same axiom schemata as PS, this is no longer the case.

§ 8: The completeness theorem

The governing idea of the demonstration of the completeness theorem is that of taking the inscriptions of a supposedly consistent theory for the model of the theory itself. This is a remarkable procedure, in which the formal assemblage articulates two functions simultaneously: the inscription of theorems and the semantic treatment of certain of its own pieces.

First of all, let us note that syntactic marks can always themselves be treated as a semantic material to the extent that their lists constitute *sets* of marks.

The universe of the model that we are going to construct is, in fact, an extension of the set of particular marks: that of the individual constants of the theory considered.

Observe that it is, in fact, possible to arbitrarily adjoin new constant marks to a mathematico-logical system: this extension is consistent so long as the initial theory is, as we can easily verify.[1]

So it is for the constants, to which we are going to assign the function of being the *elements* of a universe. The predicates will then be interpreted as follows: to the predicate *P* will correspond the subset composed of marks *a* such that *P*(*a*) is a theorem of the theory under consideration. Note that if our system were to admit of binary relations (for example), then to a relation R would have to correspond couples of constants (*a*, *b*), whenever R(*a*, *b*) is deducible. The procedure is a general one, and does not depend on our example's restriction to predicates with only one argument.

---

1. cf. for example, Elliott Mendelson, *Introduction to Mathematical Logic*, [Princeton, Van Nostrand, 1964,] p.65.

It is here that the two functions are knotted together [*se nouent*]: $P(a)$ is *valid* if and only if $P(a)$ is *deducible*. That point of saturation between syntax and semantics governs the development of the proof, as well as certain of its paradoxical effects to which we will return.

Let us now enumerate, as before, not all the formulae of T, but all the formulae that have one free variable. Let $F_1, F_2 ..., F_n, ...$ be this enumeration. To each of these formulae, we *associate* an individual constant of the envisioned model. In this numbering, we take a few precautions with respect to the differences amongst the marks, and make free use of the possibility of adding new constants.

The essential goals of these precautions are:

1°) to avoid having the constant associated with F_n either already figure in F_n or else figure amongst the formulae $F_{n\text{-}k}$ that precede it in the list;
2°) to avoid having the constant associated with F_n figure amongst the *mathematical* axioms (axioms other than those of PS) that the theory may come to contain.

Consider now all the formulae S_n of the type:

$$S_n : \sim (\forall x)\, F_n \rightarrow \sim F_n(b/x)$$

where x is a free variable in F_n, and b is a constant associated with F_n.

We are going to construct, with the aid of the formulae S_n, an infinite sequence of extensions of the initial theory T, proceeding in the following fashion:

$T_0 = T$
$T_1 = T + S_1$
$T_2 = T + S_1 + S_2$

....

$T_n = T_{n-1} + S_n$, being: $T + S_1 + ... + S_n$

....

These theories thus adjoin, to T, axioms (the formulae S_n) in which is marked a connection, internal to the assemblage, between formulae with a single free variable and the individual constants—a connection guaranteed by the serial numbering of the pieces of the assemblage. This apparatus, in sum, is controlled by a particular labelling of the formulae with one free variable.

The principal value of this control is bound up with the following result:

If T is consistent, then so is every formula T_n.

Once again, we are going to reason by a sort of descending recursion, combined with reasoning by the absurd: we are going to show that if T_n is inconsistent, then so is T_{n-1}, and therefore, ultimately, so is T_0 (*i.e.* T).

The reader may begin by rereading the demonstration of the deduction theorem (paragraph 4 of the appendix). He or she will be convinced that its result supposes only that the theory in question contains the axioms of PS, and has no rules of deduction other than separation and generalization. In other words, given a mathematico-logical extension of PS, it is always true that if B is deducible in the theory $(T + A)$, where A is a closed formula, then $(A \rightarrow B)$ is deducible in the theory T.

Suppose that T_n were inconsistent. We could then deduce from it any formula whatsoever, including, for example, $\sim S_n$. But T_n is nothing other than $(T_{n-1} + S_n)$. The deduction theorem would therefore let us say that $(S_n \rightarrow \sim S_n)$ is a theorem of T_{n-1}.

T_{n-1} being an extension of T, and thus of PS, we have the deduction:

$$\vdash (S_n \rightarrow \sim S_n) \rightarrow \sim S_n \quad \text{Axiom 4}$$
$$\vdash \sim S_n \quad \text{Separation}$$

And so, replacing S_n with its complete inscription, we have, in T_{n-1}, the theorem:

(1) $\vdash \sim[\sim(\forall x) F_n \rightarrow \sim F_n(b/x)]$

Here, we will admit without demonstration the following two theorem schemata, deducible by means of employing only the rule of separation and axioms 1, 2 and 5 of PS (a possible exercise):

(2) $\vdash \sim(\sim A \rightarrow \sim B) \rightarrow \sim A$
(3) $\vdash \sim(\sim A \rightarrow \sim B) \rightarrow B$

We will replace A with the formula $(\forall x)F_n$, and B with $F_n(b/x)$, b always being the constant associated with the formula F_n. We then have the following theorem of PS (and hence, of T_{n-1}, which is an extension of PS), a simple variant of schema (2) above:

(4) $\vdash \sim[\sim (\forall x)F_n \rightarrow \sim F_n(b/x)] \rightarrow \sim(\forall x)F_n$

This theorem of T_{n-1} and theorem (1) established above yield, by separation:

$$\text{(A)} \vdash \sim(\forall x)F_n$$

Now, schema (3), with the same substitutions, justifies (in T_{n-1} as always):

$$\vdash \sim[\sim(\forall x)F_n \rightarrow \sim F_n(b/x)] \rightarrow F_n(b/x)$$

And so, by separation again:

$$\text{(B)} \quad \vdash F_n(b/x)$$

We will now show that (A) and (B), theorems of T_{n-1}, imply this theory's inconsistency.

Let's take a look at a deduction of (*B*) in T_{n-1}. Throughout that deduction, we replace the constant *b* by a variable *y that does not figure in any of the formulae in the deduction*. This operation is always possible, because the list of variables is infinite, and every deduction is finite. Furthermore, this operation will not alter the deductive character of the sequence. In fact, it transforms the purely logical axioms into other axioms corresponding to the same schema (easily verified). It does not concern the mathematical axioms, because our precautions in the choice of associated constants guarantee that the constant *b* does not figure in any of these axioms. Finally, the axioms $S_1, S_2, \dots S_{n-1}$ are likewise unconcerned by this operation, and for the same reason. It is also clear that the rules of deduction remain unaffected: separation, because the substitution is uniform; generalization, because it does not concern the constant *b*, nor pertains to the variable *y* which, not figuring in the initial deduction, is nowhere quantified.

We thus obtain the following result: if there exists, in T_{n-1}, a deduction of $F_n(b/x)$, there certainly also exists one of $F_n(y/x)$.

By generalization, we then obtain, in T_{n-1}:

$$\text{(C)} \quad \vdash (\forall y)F_n(y/x)$$

But we have also demonstrated

$$\text{(B)} \quad \vdash \sim(\forall x)F_n$$

While, in T_{n-1}, we have the following deductive fragment:

$\vdash (\forall y)F_n(y/x) \rightarrow F_n$ Axiom 6

(replacing *y* with *x*, which is not bound in F_n)

$\vdash (\forall x)[(\forall y)\, F_n(y/x) \rightarrow F_n]$

$\vdash (\forall x)[(\forall y)\, F_n(y/x) \rightarrow F_n] \rightarrow [(\forall y)F_n(y/x) \rightarrow (\forall x)F_n]$

(Axiom 7: applicable here, because *x* does not figure in $(\forall y)\, F_n(y/x)$)

$\vdash (\forall y)\, F_n(y/x) \rightarrow (\forall x)F_n$ Separation

$\vdash (\forall x)F_n$ Separation (by C)

So $(\forall x)F_n$ is deducible in T_{n-1}, but so is $\sim(\forall x)F_n$ (proposition (B), above). The result is that T_{n-1} is certainly inconsistent. In effect, we have here the following deductive schema, with A being any formula whatsoever:

$\vdash (\forall x)F_n \rightarrow [\sim (\forall x)F_n \rightarrow A]$	Axiom 3
$\vdash \sim (\forall x)F_n \rightarrow A$	Separation
$\vdash A$	Separation

As A is any formula whatsoever, it is sufficiently clear that every formula is a theorem of T_{n-1}, and this is the very definition of inconsistency.

If, therefore, T_n is inconsistent, T_{n-1} is as well. By 'descent' we immediately see that $T_0 = T$ is inconsistent. Conversely, we can affirm that if T is consistent, then so is T_n, whatever n may be.

We will now call TU the theory obtained by adjoining, to the axioms of T, *all* the statements of the type S_n; or, if we like, the union theory of all the theories T_n. If T is consistent, then so is TU. Suppose that we could, in fact, deduce A and $\sim A$ in TU. The two deductions would have to be finite, and utilize only a finite number of axioms of the type S_n. They would therefore be internal to a theory T_n (that which contains the axiom S_n of the highest rank utilized in the deductions of A and $\sim A$). The T_n in which we deduce A and $\sim A$, would then be inconsistent (for reasons already indicated and reprised below), which, as we have shown, is impossible if T is not.

Now, after the Lindenbaum lemma, if T, and therefore TU, are consistent, then there exists a saturated extension of TU; let us call it TU'. Whereas TU is an extension of T, TU' is a saturated extension of T. We can work within TU' with the suturing structure that we have envisioned ($P(a)$ is valid if and only if $P(a)$ is deducible). If this structure is a model for TU', then all the axioms of TU' are valid there, and so are all the axioms of T, of which TU' is an extension.

In fact, we may immediately establish a stronger result: a closed formula of TU' is a theorem if and only if it is valid for TU; 'to be a theorem' and 'to be a valid formula in the suturing-structure' are equivalent statements.

The restriction to closed formulae is unimportant.

The reader may, in fact, show that:

— If F, in which x is free, is valid then so is $(\forall x)F$, and vice versa (utilize rule 5 and the definition of validity).

— If F, in which x is free, is a theorem, then so is $(\forall x)F$ (generalization), and vice versa (axiom 6).

We will reason recursively over the *number of logical signs* that figure in a closed formula. By logical signs, we understand: $(\forall x)$, $\sim$, $\rightarrow$.

a) If the formula contains *no* marks of this type, it is of the form $P(a)$. By the very definition of our structure, $P(a)$ is not a theorem unless $P(a)$ is valid, and vice versa.

b) Let us formulate the hypothesis of recursion: we assume that all closed formulae containing less than n logical signs are theorems if and only if they are valid for the structure. We are going to demonstrate that the same follows for a closed formula containing n logical signs.

c) Let A be one such formula. It may be written: either $\sim B$ (B possessing n-1 logical signs); or $(B \rightarrow C)$ (B and C having each no more than n-1 logical signs); or $(\forall x)B$ (B having n-1 logical signs).

1st Case: A is written $\sim B$

— If $\sim B$ is valid, B is not. After the hypothesis of recursion, B is therefore not a theorem. But TU' is saturated. Therefore $\sim B$ is a theorem.

— If $\sim B$ is not valid, B is. Therefore, B is a theorem (hypothesis of recursion), and so $\sim B$ is not. Now, TU' is supposedly consistent (because T is supposed to be), but if $\sim B$ and B are both theorems, one could, in TU' deduce any formula whatsoever C, and TU' would be inconsistent. Recall that, in effect, the sequence:

$$B \rightarrow \ (\sim B \rightarrow C)$$
$$\sim B \rightarrow C$$
$$C$$

would then be a deduction (exercise: generalization of a demonstration performed above).

In passing, let us note the equivalence, for our system PS, between the 'classical' definition of consistency (not simultaneously admitting both a statement and its negation) and the one that we have given (not having the power to deduce everything).

2nd Case: A is written $(B \rightarrow C)$

— If $(B \rightarrow C)$ is not valid, C = Fax and B = Vri (Rule 3). The hypothesis of recursion requires B to be a theorem, and C to not be one. But the saturation of TU' requires that if C is not a theorem, then $\sim C$ is. Under these conditions, $(B \rightarrow C)$ is certainly not a theorem. For, if it were, then since B would be a theorem as well, both C (by the rule of separation) and $\sim C$ would also be theorems. TU' would then be inconsistent.

— If $(B \rightarrow C)$ is valid: either C is valid, and is therefore a theorem, by the hypothesis of recursion, and since $C \rightarrow (B \rightarrow C)$ is an axiom, then $(B \rightarrow C)$ is a theorem by separation. Or else C is not valid, but then neither is B (Rule 3). It follows (hypothesis of recursion) that $\sim B$ is a theorem. We then have the deduction:

$$\sim B \rightarrow (\sim\sim B \rightarrow C) \qquad \text{(axiom 3)}$$
$$(\sim\sim B \rightarrow C) \qquad \text{(separation)}$$

We will admit, without further elaboration, that in any deduction, $\sim\sim B$ can be replaced by B (this involves various deductive manipulations on the basis of axioms 5, 2 and 1). Hence, $(B \rightarrow C)$ is a theorem.

3rd Case: A is written $(\forall x)B$

— If $(\forall x)B$ is not a theorem, $\sim (\forall x)B$ is (by saturation). Now, TU' contains all the axioms of TU, of which it is an extension, and therefore all the formulae of the type

$$\sim(\forall x)F_n \rightarrow \sim F_n(a/x),$$

where F_n is a formula with one free variable, and a is the constant 'associated' with F_n. As $(\forall x)B$ is a closed formula, the only free variable that B contains is x. Among the axioms of TU', there is therefore the formula:

$$\sim (\forall x)B \rightarrow \sim B(b/x)$$

By separation, $\sim B(b/x)$ is established as a theorem of TU', and so $B(b/x)$ is not (by consistency of TU'). Consequently (via the hypothesis of recursion) $B(b/x)$ is not valid. It follows that $(\forall x)B$ cannot be valid either (Rule 5).

— If $(\forall x)B$ is a theorem, then whatever constant we take for a, we know that $(\forall x)B \rightarrow B(a/x)$ is an axiom (cf. axiom schema 6), and therefore that $B(a/x)$ is a theorem by separation. The hypothesis of recursion then guarantees the validity of $B(a/x)$ for every a, and therefore that of $(\forall x)B$ (Rule 5).

Finally:

1) Closed formulae with zero logical signs are deducible in TU' if and only if they are valid (for the suturing-structure).

2) If all closed formulae with less than *n* logical signs are assumed to be deducible if and only if they are valid, it follows that the same holds for closed formulae with *n* signs.

Hence (by the informal use of the schema of recursion over the natural numbers) deducibility and validity (in the structure in question) are equivalent for closed formulae of TU'. In particular, the structure is a model for TU', and thus a model for T, of which TU' is an extension.

The only hypotheses made concerning T were that it is consistent (which guarantees the consistency of TU') and that its subjacent logic is the one expressed by our axioms for PS. We can therefore conclude:

A) Every consistent mathematico-logical theory that is an extension of PS admits of a model (Henkin's theorem).

From which it follows, as we have remarked in § 6:

B) The system PS permits the deduction of every purely logical formula the inscription of which is authorized by its stock of signs (Gödel's theorem).

These results are the cornerstone of all mathematical logic. Let us add a 'paradoxical' result: our model is denumerable, because its universe is composed of a list of numbered marks. Therefore:

C) Every consistent mathematico-logical theory that is an extension of PS admits of a denumerable model (Löwenheim-Skolem theorem).

And so even a formalized theory aiming to inscribe the structure of non-denumerable mathematical domains (like the points in a line, for example) also admits of denumerable models.

This is to say that no formal apparatus escapes the necessary capacity to inscribe its proper finitude: the discrete materiality of marks deploying the process of inscription at the heart of the apparatus. An experimental assemblage is always at the same time an experimentation in assemblage.

[Supplement]

Syntax

a) *Alphabet*

— individual constants: $a, b, c, a' b', c', ...$
— individual variables: $x, y, z, x', y', z', ...$
— predicative constants: $P, Q, R, P', Q', R', ...$
— logical connectives: negation: $\sim$; implication: $\rightarrow$.
— quantifiers: universal: $\forall$; existential: $\exists$.

b) *Rules of Formation*

— $P(a)$, $P(x)$, etc. are well-formed expressions;
— if A and B are well-formed expressions, then so are $\sim A$, $(A \rightarrow B)$, $(\forall x)A$, and $(\exists x)A$ (if x is free in A and A is well-formed).

c) *Rules of Deduction*

If A and B are well-formed expressions, and the sign $\vdash$ indicates that a formula that follows has already been deduced, we have the following deductive schemata:

Generalization $\dfrac{\vdash \quad A}{\vdash (\forall x)\, A}$ Separation $\dfrac{\begin{array}{l}\vdash (A \rightarrow B) \\ \vdash A\end{array}}{\vdash \qquad B}$

d) *Axioms*, both logical (valid in every structure), and mathematical (characteristic of the formal theory under consideration).

d') *Example*: Axiomatic of the Purely Logical System PS (cf. Appendix)

Ax 1: $A \rightarrow (B \rightarrow A)$

Ax 2: $[A \rightarrow (C \rightarrow D)] \rightarrow [(A \rightarrow C) \rightarrow (A \rightarrow \mathrm{D})]$

Ax 3: $A \rightarrow (\sim A \rightarrow B)$

Ax 4: $(A \to \sim A) \to \sim A$

Ax 5: $\sim\sim A \to A$

Ax 6: $(\forall x)A \to A(f/x)$, where x is free in A, and where f is either a constant or else an variable unbound in the part of A in which x is free.

Ax 7: $(\forall x)(A \to B) \to [A \to (\forall x)\, B]$, so long as x is not free in A.

e) *A Few Definitions and Inscriptions*

— A variable in a well-formed expression is said to be *free* if it does not fall within the scope of a quantifier. Otherwise it is *bound*. *E.g.*: In the formula $(\exists x)(P(y) \to Q(x))$, y is free and x bound.

— A formula is *closed* if it contains no free variables. Otherwise, it is *open*.

— $A(f/x)$ designates the formula obtained by replacing, in the formula A, the [free?] variable x by the mark f (an individual constant or a variable).

— If a formula A contains the free variables $x, y, z, \ldots$, a *closed instance* of A is a formula of the type $A(a/x)(b/y)(c/z)\ldots$, in which all the free variables of A are replaced by constants.

Semantics

a) *Structure*

— A set **V**, called the universe, whose elements are noted *u*, v, w, ... so that $u \in$ **V**.

— A family of (possibly empty) subsets of **V**, noted [pV], [qV], [rV], ...

— Two marks: Vri and Fax.

b) *Interpretation in a Given Structure*

— A function *f*, which: assigns an element of the universe **V** to each individual constant of the (syntactic) system, so that we have, for example, $f(a) = u$; assigns a subset of the family defining the structure to each predicative constant of the system. For example: $f(P)$ = [pV].

c) *Evaluation of Closed Formulae for a Given Structure*

Rule 1: $P(a)$ = Vri if and only if $f(a) \in f(P)$ (for example, if $u \in$ [pV]). Otherwise, $P(a)$ = Fax.

Rule 2: $\sim A$ = Vri if and only if A = Fax. Otherwise, $\sim A$ = Fax.

Rule 3: $(A \rightarrow B)$ = Fax if and only if A = Vri and B = Fax. In every other case, $(A \rightarrow B)$ = Vri Vri if and only if there exists at least one individual constant a such that $A(a/x)$ = Vri. Otherwise, $(\exists x)A$ = Fax.

Rule 4: $(\forall x)A$ = Vri if and only if, for every individual constant a, we have $A(a/x)$ = Vri. Otherwise, $(\forall x)A$ = Fax.

d) *Validity*

A formula A of a formal system is *valid for a structure* if, for every closed instance A of A', we have A' = Vri.

e) *Model*

A STRUCTURE IS A MODEL FOR A FORMAL SYSTEM IF ALL THE AXIOMS OF THE SYSTEM ARE VALID FOR THAT STRUCTURE.

An Interview with Alain Badiou

The Concept of Model, Forty Years Later: An Interview with Alain Badiou[1]

Tzuchien Tho:

Perhaps we might begin with some biographical questions. In the foreword of the book, written by the *Théorie* collective, the editors remarked that your text remained a bit too 'theoreticist' (*théoriciste*). It's a critique which is a little difficult to understand outside of the situation of May '68. I cite:

> Even today, this text's somewhat 'theoreticist' accents hearken back to a bygone conjuncture. The struggle, even when ideological, demands an altogether different style of working and a combativeness both lucid and correct [*juste*]. It is no longer a question of taking aim at a target without striking it (CM 3).

How do you view this judgment today? Does the question of 'theoreticism' still hold the same importance in the relation between philosophy and politics as it did in those years?

Alain Badiou:

The Concept of Model was a conference given... on May '68 (laughs). Thus it is very particular. The publication came in 1969. The years from '68 to '69 were perhaps the most activist, the most militant, the most revolutionary of the last fifty years of French history. For my part, I was completely involved during this period. Thus our problem concerned that which was

1. Interview conducted on 7 June 2007 at L'École Normale Supérieure, 29 Rue D'Ulm. The interview was conducted and translated by Tzuchien Tho. The questions were prepared by Zachary Luke Fraser and Tzuchien Tho. Many thanks to Bruno Besana and Oliver Feltham for help with the translation and for pointing out key passages; to Alain Badiou for his energy and generosity.

immediately militant, with organization, with interventions in factories, with interventions in housing projects. Clearly *The Concept of Model*, was something very theoretical, very abstract, etc. The judgment was thus a judgment of the times, a judgment of 1969, a judgment that was made in a period where the accusation of 'theoreticism' was frequent, very present, and concerned every activity deemed too theoretical, too formalist or too abstract — such as the very things from the immediately preceding years of '66 and '67. Evidently the question of theoreticism does not have the same importance today, but I would say that the relation between philosophy and politics today, or the question (of the role) of theory has once again become very important because the concrete situation has become very difficult and mixed. In those years we had great hope, truly massive, in the situation.

TT:

But during the publication, were you in agreement with the judgment?

AB:

Absolutely, yes. I was totally in agreement with the judgment. In fact, it was a judgment that was made with me, not against me. We were all in agreement, in 1969, to publish this judgment when we published the text. We were, on the one hand, saying that this was an important document; but, on the other hand, we were saying 'but today, we have to do other things'.

TT:

Given that the course itself was given between April and May '68, how did it go? Who attended those classes? How do you think it was received in that context and what was the reception of the text itself in the years after its publication? Was it used as a textbook for the introduction of formal logic or model logic?

AB:

The course was given at *L'École Normale Supérieure* in Dusanne Hall.[2] It was part of the group of courses, organized by Althusser, called 'Philosophy Courses for Scientists'. Apart from my text, Althusser's contribution to these courses had been published and I believe those of the other courses.[3]

---

2. *Salle Dusanne* is located on 45 Rue D'Ulm, Paris, France.

3. Of the five volumes planned, only three were published, Althusser's introduction, Fichant

TT:

I thought that there were only two texts available, *The Concept of Model* and the text on the history of science by Fichant and Pêcheux.

AB:

Actually, the introduction to the 'Philosophy Course for Scientists' by Althusser himself was also available. In any case, the course itself was a great public and intellectual space. Dusanne Hall was full. Julia Kristeva, Philippe Sollers, many others in the Paris area including almost all the young *Normaliens*, Etienne Balibar, Macherey, all of them were in attendance.

TT:

How many people could fit in the hall?

AB:

I think that at the time Dusanne Hall was not yet transformed into a projection room as it is now, so it was just a great hall with chairs everywhere, a bit like Lacan's seminars. I believe that there were perhaps three or four hundred people present. We had planned for two meetings and the first meeting was a great success. After that the second meeting was not held, because there was a strike at the university (Laughs).[4] May '68 had cut the course in two and thus the context was complex. Having said that, you know, the book was a great success. It sold close to twenty thousand copies. It was a book that, even during the years '68 and '69, was widely used, much cited and frequently mentioned. Also, as you said, it was used as a textbook and an introduction by a large number of people.

TT:

Turning to the text itself, it is interesting to see that many thinkers you mention in this text have played a less important role in your later work. If Althusser, Lacan and Maoism have constantly returned throughout your work, it seems, by contrast, that one does not often hear you speak about the epistemological tradition in the vein of Bachelard, Canguilhem

---

and Pêcheux's *Sur l'histoire des Sciences* and of course, Badiou's *Le Concept de Modèle*. The two missing texts are Macherey and Balibar's *Expérience et Expérimentation* and *Conclusion Provisoire*. These texts were published or were planned to be published by the *François Maspero* publishing house.

4. In the *Avertissement* to *Le Concept de Modèle*, Althusser, referring to instalment II of Badiou's presentation, says it was 'happily interrupted', p. 7.

and others like them or, to add to this list, figures like Michel Serres with whom you did the interview in *Modèle et Structure*.[5] Also, what was the role of the work of Albert Lautman and the Bourbaki group in your work in those years?[6]

AB:

It is true that the epistemological references were very explicit in *The Concept of Model*. I think it is because these were the references from my years of education. My formative years were very contradictory. On the one hand, we had Sartre and Existentialism and, on the other, we had mathematics and the field of epistemology. Also, in those years of my education, Canguilhem and Bachelard were very important, as was Althusser. I still consider these references quite important today, even if they are no longer explicitly cited. As references from my years of education, I continue to think of Bachelard and Canguilhem as important French thinkers even if I don't cite or mention them. In the work during those years, the Bourbaki group was extremely important. It was where I studied set theory and formal mathematics. I was less familiar with Lautman because his texts were practically impossible to find. It was much later that I rediscovered Lautman. He was practically unknown in those years because the texts were unpublished, although there was an edition of some of the texts that appeared in the seventies.

5. *Modèle et Structure* is a television film directed by Jean Fléchet and produced by Dina Dreyfus under the series name *Philosophie*. It was made in association with the Centre Nationale de Documentation Pédagogique in 1968. The series served as one of the materials to be studied for the philosophy section of the university entrance exam (*baccalauréat*). These particular episodes, three in total that comprise the film, were published in 1968 and were recordings of an extended interview between Michel Serres and Alain Badiou. Badiou refers to the many examples used in the film in *The Concept of Model*. Despite its availability during the period when Badiou wrote the notes to *The Concept of Model*, the transcription of the film, in its original form, is no longer retrievable. A new transcription and translation into English is currently in preparation and will appear in the journal *Cosmos and History*.

6. Nicolas Bourbaki was an allonym for a group of mathematicians based in France in the 1930s and 40s who worked to reorganize all of mathematics on the basis of set theory. They are cited by Badiou in *The Concept of Model*. It is also interesting to note that the Bourbaki group has an official office at *L'École Normale Supérieure*. Albert Lautman was certainly associated with Bourbaki but was not a member. His work in philosophy of mathematics concerned the question of the reality and conceptual structure of mathematics. Badiou mentions his 'Platonism' in the introduction of *Metapolitics*, together with Lautman's contemporaries, Jean Cavaillès and Georges Canguilhem. Alain Badiou, *Metapolitics*, trans. Jason Barker, London, Verso, 2005, pp. 2-8. Badiou also discusses the continuity of his work and Lautman's in *Briefings on Existence: A short treatise on transitory ontology*, trans. Norman Madarasz, Albany, SUNY Press, 2006, p. 60.

TT:

You mean the one by the publishing house 10/18.[7]

AB:

Yes, that's it, the publication by 10/18 appeared in the seventies and now there is another edition by Vrin.

TT:

And Cavaillès, what importance did he hold for you?[8]

AB:

Cavaillès was better known and studied in those days but, to speak honestly, Cavaillès did not have an impact on me. There is significant disagreement between us. I cannot really develop this point here but I think that the account of mathematics, of set theory, of the history of formalism and its interpretation that I give is very different from Cavaillès. I consider him a great mind, so the problem is not there. However, these fundamentally different views are not frequently cited. With Michel Serres it is something of a different matter. He is someone that I know personally and was my interlocutor in the film (*Modèle et Structure*). I was never really close to him and even less so in the following years.

TT:

But there seemed to be a significant agreement that developed in the film as to the question of the differences between theoretical models and structures.

AB:

Absolutely, on that point we agreed. In those years, he had only written the big book on Leibniz which was something I found interesting.[9] But

---

7. The publishing house 10/18 published a collection of Albert Lautman's texts in 1977 under the title *L'Unité des sciences mathématiques et autres écrits* edited by Maurice Loi. This text has since been out of print. In 2006, a new edition was published by Vrin as *Les mathématiques, les idées et le réel physique*, Paris, Libraire Philosophique, J. Vrin, 2006.

8. Jean Cavaillès was a philosopher of mathematics active between the 1930s and 1940s. Much of his research concerned axiomatics, mathematical logic and the history of science. Shot by the Gestapo in 1944, he is highly regarded as one of the leading figures of the WWII French resistance. Badiou speaks of Cavaillès in *Metapolitics* as one of the 'resistant philosophers', *Metapolitics*, p. 7.

9. Michel Serres, *Le système de Leibniz et ses modèles mathématiques*, 2 vols., Paris, Presses Universitaires de France, 1968.

after the seventies or eighties, his work became further and further removed from the horizon of my own.

TT:

I am curious about something very interesting that Serres said in the program. He used the word 'fidelity' in relation to Leibniz's usage of the term. It seems like there is an echo here with your (later) work. Is there a real connection?

AB:

No, it is a coincidence. In reality, Serres has developed his work in a totally different way. But at the same time, I went to his classes at *École Normale* when he was teaching some courses on Leibniz. I was preparing for the *agrégation* and there were some texts on Leibniz that I was studying. I had a high esteem for him.

TT:

You speak today about the orientation of thought.[10] How did you orient the relations between mathematics, ontology and logic in those years? It seems that reflecting on the relation between mathematics and logic has been central to your work since then. You write that:

> The surest criterion amounts to saying that an axiom is *logical* if it is valid for *every* structure, and mathematical otherwise. A mathematical axiom, valid only in particular structures, marks its formal identity by debarring others through its semantic powers. Logic, reflected semantically, is the system of the structural as such; mathematics, as Bourbaki says, is the theory of *species of structure* (CM 35).

In using a distinction that comes a bit later, mathematical axioms are ontological decisions on pure multiples. On the contrary, logic is that which concerns existences or appearances.[11] Is the original distinction that you make between logic and mathematics something that anticipates the ontological positions that you lay out in *Being and Event*? If one holds

10. This was the subject of Badiou's seminar at *L'École Normale Supérieure* during the academic year 2006 to 2007, entitled *Comment s'orienter existence*? The title plays on a 1785 text by Immanuel Kant, 'What does it mean to orient oneself in thinking?'

11. The distinction between mathematics and logic is the distinction between the science or formalization of pure being and appearance respectively, see Alain Badiou, *Logiques des mondes*, Paris, Éditions du Seuil, 2006, pp. 44-49.

the distinction of logic and mathematics to be the distinction between existence and being, how can we orient ourselves vis-a-vis the relation between the global and local in *The Concept of Model*? I also recall that, in *Briefings on Existence*, you argued that this precise distinction is not sufficient to liberate us from a linguistic determination.[12]

AB:

I think that the response to the relation between logic and mathematics in *The Concept of Model* was the beginning of my reflections on these things. There was no other starting place! We might say that logic was the thought of what was structural and mathematics was the thought of the system structured. One should be able to make a distinction between the two. Here we have something that anticipates my later work. At the same time, it is likely that, in my subsequent development, this distinction had become very insufficient for me except as a formal point of departure. From the moment that I gave a properly philosophical interpretation, that is, more than an ontological or phenomenological interpretation of this distinction, it was necessary to add a supplementary element. It was no longer possible to consider this (question) solely on the basis of a formal distinction. It is true that with respect to logic and its axioms, the interpretation is quite vast, with many models or many possible models, while in mathematics, its axioms or structures, we have something much more specific. But in the last instance, in my subsequent interpretation, mathematics is the ontology of the pure multiple, while logic definitively describes possible worlds, describes the formal structure of a possible world.[13] We have introduced something here which is no longer something between pure formal universality and existence of particular models but rather something between a general theory of multiplicity and a local theory of multiplicity. I moved then, from proposing a purely formal theory to a more topological question of being-there and the locality of being in its becoming-indeterminate.

12. Badiou, *Briefings on Existence*, p. 112. In this passage in *Briefings on Existence*, Badiou directly criticizes the insufficiency of the definition that he had originally put forth in *The Concept of Model*.

13. Not to be confused with David Lewis' possible worlds, the account of the structure of a 'world' is the heart of Badiou's recent book, *Logiques des mondes*, where, by analysing what he calls a 'transcendental', 'objects', 'relations', 'points' and 'bodies' come to form the structure of the logic of existence. A focused explanation of a paradigm of a world can be found in his chapter on what he calls 'classical worlds', see *Logiques des mondes*, pp. 195-202.

TT:

Has there been a reversal in your formulation of the distinction? It seems that mathematics has become something which carries less specificity, while being-there has become connected with logic rather than mathematics.

AB:

One should be careful. In *The Concept of Model*, where the criteria were formal, the question was correlated to the universal quantifier.[14] Here, if we say, 'for all x, something', then it was logical if the proposition and the universal quantifier were valid for every possible model, and mathematical if the universal quantifier was available for determinate models. This gives the impression that logic is much larger and mathematics more particular. The reversal consists in, one might say, that today mathematics touches being, strictly speaking, and logic does not touch anything other than the localization of being. But that does not mean that mathematics is narrower, because every localization finally includes the presence of pure multiplicity as a matter of localization. It is now the case that the criteria for the local and the global are no longer the same. I would say this, rather than positing a simple reversal. The criteria for the local and the global are no longer the same for the reason that, in *The Concept of Model*, the criteria for local and global were purely formal; there was an extension of the value of the universal quantifier. It is in fact much more complicated than that. The question today of the local and the global is a localization of the global itself.[15] The world is a localization of what exactly? It is the possibility of thinking being as such in a determined world. Thus it is rather a modification of the relation between the local and the

14. Quantificational logic continues to be important for Badiou. It is especially interesting to compare the discussions of quantification in *The Concept of Model* and *Logiques des mondes*. See *Logiques des mondes*, pp. 191-194.

15. Among the stakes of *Logiques des mondes* is its focus on locality. On one hand, Badiou maintains that the science of being as such is ontology as mathematics, while the local investigations of being or worlds comprise 'logic' in the recent text. This distinction proves to be a difficult and often confusing one. To add to the confusion, in the early texts, including *The Concept of Model*, Badiou marks a clear distinction between mathematics and logic in an altogether different way. Here Badiou clarifies and distinguishes his early approach from his later approaches with an explication of the manner in which the global and local itself are recognized. For an account of the global and the local in *Being and Event*, see meditations seven and eight, pp. 81-103. For a recent account of the distinction between logic and mathematics, see chapter eight of *Briefings on Existence*, pp. 107-113. For an account of the global and local in *Logiques des mondes* see book III, section one and three, pp. 211-244, pp.257-282.

global rather than an inversion of the role of the extension of the quantifier. This is what I mean when I say that this delimitation is not sufficient to liberate us from linguistic determination. It is because it is a delimitation that continues to function in a purely semantic manner, that is to say, in the manner of the interpretation of the universal quantifier.

TT:

The topic of formalism has long been central to your work, and has recently been revitalized in *The Century* and *Logiques des mondes*. *The Century* concludes with the directive to overcome the 'animal humanism' [*l'humanisme animal*] that characterizes our current ideology by way of a 'formalized in-humanism' [*in-humanisme formalisé*].[16] *Logiques des mondes* defines the subject as being, in general, '*un formalisme porté par un corps*'.[17] Among other things, *The Concept of Model* offers us an extended analysis of what mathematical formalization is, shedding light on the way in which this operation is thoroughly bound up with mathematics' relation to its own historicity. Formalization, in *The Concept of Model* and other early texts, however, is understood as an important dimension of the a-subjective process that is mathematics, whereas the concept later reappears as a dimension of the subject itself.

My questions here are very general. First of all: what has changed in your understanding of the subject and formalism, such that they have ceased to be mutually exclusive and come to be implicated in one another? Second: how far can we generalize the key theses in *The Concept of Model* on formalism—are they specific to mathematics, or can they be brought to bear on the other modalities of subjective formalism addressed in your recent work? What are the essential differences between, say, mathematical, artistic, and political formalism?

AB:

This is a very just, very good question. I think you have posed a central question. This is a question of the relation between formalism, that is, form, and the theory of the subject.[18] It is clear that in the first part of my work, the general orientation was one of separation, of delimitation,

16. Alain Badiou, *The Century*, trans. Alberto Toscano, Polity, Cambridge, 2007, pp.177-78.

17. That is, 'A formalism carried by a body', *Logiques des mondes*, p. 593.

18. Badiou develops the relation between form and formalism in *Being and Event* especially with respect to his notion of subjectivity in meditation twenty-two, 'The form-multiple of intervention'. *Being and Event*, pp. 223-231.

between formalism, considered as completely objective activity, perhaps even as objectivity itself, and the subject, which I interpreted, in those days, a bit like Althusser, that is, as an ideological given (I think because I was still an Althusserian at the time even if he later talked about the subject as ideological interpellation). I think in that period I was freeing myself from Sartre, existentialism and phenomenology, but unfortunately it was not a focused or deliberate effort and thus the subject was rejected as ideology. This brought me to a critique of Lacan that we see in *Marque et Manque*, an article published around the same time, where I said that the thesis of the suture is a thesis which does not permit us to account for mathematical formalism, which I considered as principally non-subjective. This is to say that mathematical formalism is neither sutured nor subject but, in fact, non-subjective.[19] Here I was at a preliminary point of departure. I would say that what first seduced me in my mathematical education was the non-subjective, the making possible of a capacity to think outside of all intentionality and subjectivity. I reconsidered this point when I came to understand that, even if it is not right to consider formalism as something constituted as intentional subjectivity, it was necessary to take and maintain some aspects of subjectivity in the elements of formalism itself. I am not saying this for political reasons but I remain convinced that every philosophy that eliminates the category of the subject becomes unable to serve a political process. That is not to say that a subject should be identified from the outset as the working class or the like. Certainly, there is subjectivity in politics and there is subjectivity in art and subjectivity in love and subjectivity in science itself. As such, I attempted to completely rethink the relation between formalization and subjectivity.

To answer the first question, I attempted to recommence, as one should always do, with the theme of separation, radically and absolutely. From *Théorie du Sujet* and onwards in the 1980s, I began again to re-knot these terms in a different way.[20] But that is not to say that I abandoned the primacy of formalization because it remained a thought. I found it necessary to re-knot this formalism with the figure of the subject. But, from that, what do I conserve of the separation or what do I still hold of the separation? I think the separation I maintain is the idea that the relation between the subject and formalism is on the side of formalism and not

19. Badiou, 'Marque et Manque'.

20. Alain Badiou, *Théorie du sujet*, Paris, Éditions du Seuil, 1982.

on the side of the subject. In the rigorous examination of formalization, one can dispose or place the subject, ultimately, as an effect and not as a cause. Thus, finally, an event is that which renders possible a new formalism for such a relation. The subject will be the subject of the formalism or with respect to this formalization. Where there is an effect of puncture in the particular underlying structure, the subject will be defined as a new process of formalization. It is a bit messy, but there you go. If we want to think the subject, one should begin by thinking through formalization.

On the second point, formalization, in its essence, is not only mathematical or logical. While mathematical and logical formalization is a paradigm for formalization, formalization is not identical with this. The question of knowing what is, or how one should analyse the formalization of artistic or political formalization comes down to a question of analysing those very sequences. If we have an event that makes possible a new formalization, we should study this possibility of formalization for itself. There, the principles of formalization, even if they can be comparable to or analysable with mathematics, would not be identical. Definitively, the study of the different types of generic procedures is truly the study of the different types of formalization. You know, as I described in *Logiques des mondes*, if we are to begin a study of a particular formalization, it would be necessary to return to the position of the difference between formalization and the evental *énoncé*, the stakes are always there.[21] Well, what follows from this are the enquiries particular to the process. Thus the general schema is that every event is an opening of a new possibility of formalization, carried forth by a new body. This new body always supports the formalization with respect to formal articulations.[22] This is maintained for every truth procedure, and the study of particular formalizations will be the study of a regional world.

TT:

As is well known, mathematics has always held a privileged place in your thought, from your earliest published works to the most recent. Beyond this rather general interest in mathematics, however, model theory—the topic of your first book—seems to stand out as of utmost importance. From *The Concept of Model* to *Being and Event*, several of the canonical

21. Badiou discusses the relation between formalization, subjects and the types of truths in evental announcements (*énoncé*) in *Logiques des mondes*. A schematic presentation is given in book I, section 8, pp. 81-87.

22. A theory of bodies is developed in book seven of *Logiques des mondes*, pp. 471-526.

mathematicians in your work are those whose greatest breakthroughs have been in the theory of models—Gödel and Cohen leap to mind here.

In the work of these mathematicians, we see model theory employed in a manner that strikes a profound resonance with several of your own writings; it combines, in an almost paradoxical fashion, an attentive examination of the effects of formalization with a sort of underlying Platonism which aims to locate the precise points where a given formalized theory transcends itself, or opens onto a point of undecidability. I say that this combination seems paradoxical because we find, often together in the same texts, a commitment to the (formalistic) identity or immanence of mathematical thought and its mode of expression, but also the espousal of the Platonistic idea that the reality accessed by mathematical thought transcends its mode of expression. Without the former conviction, one could not seriously take, say, a denumerable model as offering a genuine interpretation of the axioms of set theory, even if Löwenheim's theorem proves to us that such an interpretation is possible. (Bernays dismissed the importance of Cohen's findings along these lines.[23]) Without the latter conviction, neither Cohen nor Gödel would have grounds to see several of their results (the incompleteness theorems, the independence of the continuum hypothesis, etc.) as any indication of the limitations of mathematical thought.

My question is: what exactly is the importance of the theory of models in the context of your general understanding of mathematics? Is there a connection between your early theses on model theory and the formalistic Platonism that you espouse in your later works (notably in *Briefings on Existence*)?

AB:

Evidently, the concept of model is absolutely central because it is the heart of what follows from formalization. The fact that I began with the concept of model is not a coincidence. It turned out well because, at base, a model is a concept I investigated in an attempt to focus on what one could call the dialectic of formalization. I call it the dialectic of formalization due to the fact that every creation of thought is in reality a creation

23. See Paul Bernays, 'What Do Some Recent Results in Set Theory Suggest?' in Imre Lakatos (ed.), *Problems in the Philosophy of Mathematics*, Amsterdam, North Holland Publishing Company, 1967, p. 82.

of a new formalization and at the same time this new formalization establishes a relation or takes part in an interaction with the particularity of what we are trying to express. In this case, we determine the formalization as a universality, but it is ultimately a particularity that carries universality in the model. Because, at base, we can say that, even if we take, for example, a painting by Picasso, that is, if we are taking a cubist painting by Picasso or by Braque in 1913, we find the creation of a possibility of a new type of pictorial formalization.[24] That is to say, it renders possible a way to formalize in the space of painting something that was previously unacceptable. On the other hand, it realizes itself in a particular context, with respect to the materials used or in the sorts of cultural references that render the painting a particular painting. It is a model. Picasso's work is a model of this possibility of formalization. It is not that the formalization is drawn abstractly from thin air. It is rather something that was realized in a particular time and place. And thus the model, since my earliest reflections, has been something that assumes the particularity or the singularity of a region of being or of a world, and at the same time, raises the universality of a possible new formalization. This is why it is effectively a concept that describes the dialectic of formalization. That is to say that it is a dialectic of the truth procedure as such.

It is entirely situated then, in what at times can be called 'situations' and other times 'worlds' and, at the same time, it transcends its situation because it proposes a new type of formalization which has the power to be summoned up in the history that follows. Thus, if we return to the dimension of mathematics, we will see that we have been given completely bare, as one might say, a particularity with least possible particularity. Thus while mathematics is a particularity just like everything else, mathematics will be the least particular or that which carries with it the least particularity.

TT:

Right, as you said in the Duchamp lecture, 'the least particular particularity'.[25]

24. In his lengthy introduction to *Logiques des mondes*, Badiou takes Picasso's *Deux chevaux traînant un cheval tué* (1929) and *Homme tenant deux chevaux* (1939) as artistic examples of the logical treatments of truth procedures in worlds. In this context, he relates these works to the prehistoric grotto paintings of horses at Chauvet-Pont-d'Arc, Ardèche, see *Logiques des mondes*, pp. 25-29.

25. Badiou used this expression during a lecture on the work of Marcel Duchamp at L'École Normale Supérieure, March 3 2007, seminar series of the *Centre Internationale d'Etude de la*

AB:

Exactly. Thus, what we can see here is that the model is that which permits the study, on one hand, of the power of formalization, but also on the other hand, at its limits, that it becomes something that permits us to arrive at a dialectical point, the most concentrated point. These are specific points or certain times, at which the infinite power of formalization and its limitations are irreducible and present a point of undecidability distinct from the others. Thus, what interests me in particular is something that in fact supports my peculiar Platonism. I should say that Platonism, in the end, is the knowledge of ideality. But this is also the knowledge that we have access to ideality only through that which participates in ideality. The great problem of Platonism is not really the distinction between the intelligible and the sensible, but the understanding that sensible things participate in the intelligible. What interests me in Plato is participation and what would otherwise be very obscure becomes very clear. It is central to note that, in the end, the eternal truths as well as ideas would be nothing if they were incapable of being accessed from what is given in the sensible. We talk about the idea of the table, to take a classical example, that there would be no idea of a table if there were not a table. We would have no idea of the idea of the table if there were not tables. We could take other examples with things that are on a level a bit more technically formulated. The model is thus that which allows us to conceive formalization; conceived after the fact, given mathematical inventions are not simply formal inventions but rather an invention of models. It is that which permits us to access formalization or to access the universality of things, at the same time it permits us to determine the particular point of limitations. To put it more directly, the model is that which allows us to think through participation.

TT:

So it seems that there is an interesting history here, your personal history. As you said, you discovered Lautman after this period, but for Lautman the question of Platonism was exactly that of *metaxu* or participation.[26]

Philosophie Française Contemporaine (CIEPFC).

26. See Albert Lautman, *De la réalité inhérente aux théories mathématiques* and *Essai sur les notions de structure et d'existence en mathématiques* for short but precise discussions of platonic participation in modern mathematics. Lautman, *Les mathématiques, les idées et le réel physique*, pp. 65-69, 125-223.

AB:

Absolutely, that is right. Lautman directly expresses this point. As he put it, in a certain sense, mathematics is the *model* of the dialectic; but for Plato, mathematics was an *introduction* to dialectics. The Lautmanian interpretation centres on the dialectic of ideas in the history of mathematics to which we finally gain access with Gödel.

TT:

So it seems to me that there are two sides of Platonism here, of which we have already spoken. On the one hand, with Gödel, we have a sort of classical Platonism or, as it were, an Anglo-Saxon version; and on the other hand, there is something else that you seem to be closer to, the Platonism of Lautman.[27]

AB:

Yes, absolutely. I think that while Gödel's Platonism is a Platonism of ideal objects, of formalization as the construction of ideal objects, Lautman's Platonism is a Platonism of participation. What I would say is that we have a Platonism that's a bit too dogmatic versus a dialectical Platonism (or something like that). Gödel himself struggled against the American trend of empiricism and against this American empiricism; he said, no, mathematical objects exist in themselves. Thus he maintained an over-idealized Platonism against this trend of empiricism. For us, in France, we had much less of this confrontation with empiricism; instead the question of dialectic has always been central. Thus, there is more development in my work apropos of mathematics that engages in a dialectical interpretation of Platonism. As such, the heart of dialectics in Plato is the question of participation. It is not so much the distinction between the sensible and the intelligible. Rather, I found it more useful to think through the problem of the intersections between the sensible and intelligible that Plato called participation, something which is finally formally realized in the concept of model. A model is developed from a particular world, and it participates through the idea of formalization.

27. For Badiou's discussion of his opposition to a standard Anglo-analytic formulation of Platonism in mathematics as represented by Fraenkel and Bar-Hillel—a camp in which Gödel is often included—see *Briefings on Existence*, pp. 89-99 (also published as 'Platonism and Mathematical Ontology', trans. and ed. Ray Brassier and Alberto Toscano, *Theoretical Writings*, London, Continuum, 2002, pp. 49-58). However, in this text Badiou nuances his interpretation by distinguishing Gödel from a standard expression of mathematical Platonism.

TT:

In *The Concept of Model*, you make a thorough case against what you call the bourgeois-epistemological versions of the category of model, both in the 'vulgar' form you encounter in Lévi-Strauss and others, and the sophisticated form you find in positivist epistemologies such as Carnap's. The fault you find with the former is that, by envisioning science as the confrontation between formal models and empirical reality, it amounts to little more than the importation of the ideological opposition between thought and reality, or culture and nature, into the philosophy of science; an importation that more often than not operates by crudely analogical means.

As an example, you cite a few passages from Von Neumann and Morgenstern's *Theory of Games and Economic Behaviour*, where the authors write that models 'must be similar to reality in those respects which are essential in the investigation at hand', and that 'similarity to reality is needed to make the operation significant'.[28] Here, we get a depiction of scientific activity that amounts to little more than 'the fabrication of a plausible image', and the historical, productive and transformative character of science is effaced (CM 16). You recognize positivist epistemology as attaining a far greater level of rigour and fidelity to the logical *concept* of model (i.e. positivism exhibits a true philosophical *category* and not a mere ideological *notion*), insofar as it both reverses the vulgar image of science and treats the empirical as the model of the formal, formulating a rigorous set of rules for the interpretation of the scientific syntax in an empirical model.

Nevertheless, one does not seriously break with the bourgeois depiction of science as an imitative activity, concerned first and foremost with drawing a correspondence between the formal and the empirical. Once again, the historicity, productivity and transformational capacity of science are ignored. Bearing this in mind, it surprises me that in *Being and Event* we find the situation of ontology (set theory) opposed to 'non-ontological situations' in a way that seems to mimic, down to the details, the bourgeois opposition between the formal and the empirical. Like Carnap, you seem to treat non-ontological situations as if they can legitimately be understood as 'models' of the ontological situation (for instance, in order for Cohen's methods to be of any use in understanding

28. John Von Neumann and Oskar Morgenstern, *Theory of Games and Economic Behaviour*, Princeton, Princeton University Press, 1953, p. 32. Cited in CM p. 16.

truth procedures, our situation must be understood as being analogous to a denumerable, transitive model for the Zermelo-Fraenkel axioms). Like Von Neumann and Morgenstern, you seem to leave the operation of correspondence between the ontological situation and its non-ontological 'outside' up to the vagaries of analogy, as when you write, at the beginning of mediation twelve in *Being and Event*:

> Set theory, considered as an adequate thinking of the pure multiple, or of the presentation of presentation, formalizes any situation whatsoever insofar as it reflects the latter's being as such; that is, the multiple of multiples which make up any presentation. If, within this framework, one wants to formalize a particular situation, then it is best to consider a set such that its characteristics — which, in the last resort, are expressible in the logic of the sign of belonging alone, ∈ — are comparable to that of the structured presentation — the situation — in question.[29]

How does *Being and Event* avoid or respond to the criticisms voiced in *The Concept of Model*?

AB:

Yes. In a certain sense, the difficulty that you point out in *Being and Event* is part of the origin of *Logiques des mondes*. You see, *Logiques des mondes* shows that the relation between a formal ontology of sets and the question of effective situations cannot be adequate to being a relation established through the form of analogy. This is true but I have never read a critique in the way you have posed it. But it is in a certain regard quite pertinent. There (in the passage quoted), I was concerned with the idea, perhaps not very well posed, that the ontology of the pure multiple can be a model of a concrete situation. In reality, this does not at all capture the real relation. While real situations are composed of pure multiplicity like everything else, there are nonetheless some different parameters. This is why the transcendental was introduced in *Logiques des mondes* as an element which localizes or which topologizes multiplicity and not as something which turns out to be a formal imitation.

What happened in *Being and Event* was that I left to one side the particularity of the situation, or, as I announced in the beginning, that I occupied myself with the ontological structure of multiplicity, an abstraction

29. Badiou, *Being and Event*, p. 130.

made full by the singularity of the situation (of mathematics) and thus I developed a theory of the ideality of the multiple and not a theory of participation. Hence I gave myself the goal in *Being and Event* to attempt a formal description of what a truth procedure could be.

I might clarify myself with respect to the critical passage here by saying that, since there is a diversity of situations as well as laws governing situations, the question might then be how these diverse situations participate in the same ontology. In *Being and Event* what interested me was to arrive at thinking the possibility of a process of truth as an exception from a general regime of the simple repetition of the multiple. If you like, I would say that, at each point of my approach to the question of the particularity of the situation in *Being and Event*, I treated them as something related to the situation as a multiplicity. Thus, to be able to say that *this* multiplicity is different from another, I had no other recourse than an analogy to set theory. The root of the problem is that when we want to pass from the theory of pure multiples to the theory of a composed world, we need resources that are not available in *Being and Event*. In *Being and Event*, we had that which one could strictly access from the structures of belonging or inclusion of the void, with a certain number of the formalizations of set theory. From this sort of approach, I gave an account of what an event is, what fidelity is, or what the consequences of the construction of generic truth are. In the end, *Being and Event* achieved its goal when it specified the ontology of truth as the being of the construction of a generic multiplicity. Certainly, this is valid for all truth procedures. This is all good, except for what concerns the particularity of each kind of truth procedure. In the last instance, the particularity of truth presupposes that we can think the particularity of a world and we could not think this particularity with the materials of *Being and Event*. In this sense we might say that the criticisms in *The Concept of Model* were an anticipation of *Logiques des mondes*. The positive dimension of the critique was realized more in *Logiques des mondes* than in *Being and Event*.

TT:

It seems it would be a confusion to say that the four procedures of truth, from different situations, are all connected to the base of ontology or the base of mathematics. This would be a misunderstanding of your work.

AB:

Yes, certainly. While we might say that the four processes of truth, in a strictly formal sense, continue to be held together as generic multiplicity, by doing so we would not have said anything about their differences. As such, we cannot base the four truth procedures in set theory. For this reason, we need to engage, at one point or another, with that which gives some indication of singularities, that is, an investigation into the question of what is a 'type' in truth procedures. This is a question that is not reducible to ontology. Ontology can only give what it can give, it cannot but be what concerns multiplicity as multiplicity. However, when a break is situated in a world, in a certain historical shift or an interruption as a particular event, we are in a different register and it will be a different question.

TT:

So perhaps we can reformulate this general confusion as a confusion concerning the terms situation, presentation, multiple, and now... a world.

AB:

Yes. (Laughs)... and now a world.

TT:

Situation, presentation, multiple, world.

AB:

I should clarify this point. They are not at all identical. Presentation is a particular qualification at the level of situation; presentation is a synonym of belonging (*appartenance*).[30] In a situation, there is not only presentation but also representation because we should not forget that there is also a state of the situation. They are not at all the same, presentation and situation. On the other hand, the entire effort of *Logiques des mondes* has been to ask another question. Why is it that I speak about worlds? A world is not reducible to a situation because, if we look at *Being and Event*, a situation is a multiple. That is all. As such, with respect to a world, we have something that is already much more complicated. First, there is not simply a multiplicity, but there is an inaccessible infinity of multiplicity and there is also a transcendental map of connections and relations.[31] You

30. This is a central thesis of *Being and Event*, meditations three and seven, pp. 38-48, 81-92.

31. In *Logiques des mondes*, Badiou gives an account of these terms in book four, section one,

are absolutely right to point this out. Situation, presentation, multiple and world are concepts that should remain distinct.

TT:

In *Marque et manque*, a text published around the same time as *The Concept of Model*, you level what is, in my view, a decisive critique of Jacques-Alain Miller's attempt to connect the psychoanalytic concept of 'suture' with the mathematical zero.[32] The main thrust of your argument is that mathematics lacks nothing that it does not produce, that it organizes these lacks according to a process of stratification, and therefore never encounters the sort of uncontrolled lack such as Miller posits in 'Suture'. In *Being and Event*, however, the concept of suture returns, and you say that the empty set (Ø) is, in truth, set theory's 'suture to being'. What is unsettling about this return is that nowhere do you provide a defence against your earlier attacks on this concept, neither in *Being and Event*, nor in your brief discussion of Miller's text in *Le Nombre et les nombres*, despite the prominence of the notion of ontological suture in both books.[33] I am aware that much has changed in your understanding of mathematics between *Marque et manque* and *Being and Event* — mathematics, for instance, goes from being understood as a process, even a 'psychosis', without a subject, to being at once the science of being and a subjective procedure par excellence — but it is unclear how these changes might allow you to evade your earlier critique.[34] Is the notion of suture, as it functions in these later texts, still subject to the critique put forth in '*Marque et manque*'? If not, why not?

AB:

Yes. I understand the complexity of your question. However, the problem here is that the word suture changes its meaning between Miller's text and my usage in *Being and Event* when I discuss the suture to being. Why is there this change of meaning? In Miller's text, suture designates the point of absolute lack which accommodates the heterogeneity of the

---

entitled 'mondes et relations', pp. 319-342.

32. Jacques-Alain Miller, 'La suture (elements de la logique du signifiant)', *Cahiers pour l'analyse*, no. 1, pp. 39-51. Available in English as 'Suture (elements of the logic of the signifier)', *Screen*, 18.4, pp. 24-34.

33. Alain Badiou, *Le Nombre et les nombres*, Paris, Éditions du Seuil, 1990.

34. Badiou's use of the term 'psychosis' comes out of the critique of Jacques-Alain Miller's use of the term to describe the structure of science. Jacques-Alain Miller, 'Action de la structure', *Cahiers pour l'analyse*, no. 9, pp. 93-105.

subject. At the point where there is lack, one can discern the symptomatic of the subject. Here I would repeat my disagreement with Miller, while clarifying that this problematic does play the same role in *Being and Event*. For me suture designates a juncture between ontology and its 'object'. To hold mathematics as ontology, the very limits of being as such will be touched by the void. But, as such, the void is not the point where we discern subjective heterogeneity. The void is the point on which we found the constructible sets which allow us to unfold the characteristics of pure being. In this sense the void also represents inconsistency.[35] If we admit both that multiplicity is inconsistent and that ontology makes itself consistent in that possibility. With inconsistency (of the void), we are at the point where it is equivocally consistent and inconsistent. That is the void. Since the void is the multiplicity of the nothing, the question of knowing whether it consists or not is split by a pure mark (Ø). Suture carries its importance in my work in this way in relation to this difference in meaning. For Miller, it is a dynamic according to which all repetitions are conditioned by the marking of lack as hidden subjectivity. In *Being and Event* however, I simply provided a justification of the fact that mathematics, since it is consistent, is an ontological discipline that measures its (suture's) connection with the void. Whether the void is consistent or inconsistent is undecidable. Thus, it is both that (in the void) there is nothing that consists and which can yet be considered to consist in not consisting.

That is what I have to say on suture but I would like to return to another part of the question that was supposed in what we spoke of earlier. There is an important transformation in my work that is not really addressed in the question of suture by itself. Here, I think there has been an important transformation in my work between *Marque and manque* and *Being and Event*: between the two, I have reintroduced the category of the subject. By consequence, even in thinking of mathematics as a truth procedure, there is a sense of the subject in this procedure. In doing this, I did not put the subject on the side of the void. In what I developed elsewhere, I showed that the fundamental difference between Lacan and my position is that the central concern for mathematics in Lacan is his suggestion that the void is on the side of the subject, while for me the

35. In *Being and Event*, many of Badiou's central theses on the identity of ontology and mathematics hinge on the point that 'The void is the name of being', *Being and Event*, p. 56. Badiou gives an account for the void's linking of consistent and inconsistent multiplicity in meditation four, pp. 52-59.

void is on the side of being. So this is precisely the gap between the two senses of the word suture. On the one side, we have a suture to being and on the other, a suture of subjectivity. But at the same time it remains true that the construction of this interpretation of mathematics makes possible the existence of the subject in the mathematical procedure insofar as the subject of mathematics will not be localized in the void. Rather, the void is not at all the subject but rather the 'object' of the procedure. This is the important shift.

TT:

In *A Thousand Plateaus*, Deleuze and Guattari attempt to produce a novel philosophical category of model, one which they hope to be capable of rendering the operations of Capital intelligible. According to Deleuze and Guattari, capitalism should be understood as an 'axiomatic', which finds its 'models of realization' in western states. They write:

> Politics is by no means an apodictic science. It proceeds by experimentation, groping in the dark, injection, withdrawal, advances, retreats. The factors of decision and prediction are limited. [...] But that is just one more reason to make a connection between politics and axiomatics. For in science an axiomatic is not at all a transcendent, autonomous, and decision-making power opposed to experimentation and intuition. On the one hand, it has its gropings in the dark, experimentations, modes of intuition. Axioms being independent of each other, can they be added, and up to what point (a saturated system)? Can they be withdrawn (a 'weakened' system)? On the other hand, it is of the nature of axiomatics to come up against *so-called undecidable propositions*, to confront *necessarily higher powers* that it cannot master. Finally, axiomatics does not constitute the cutting edge of science; it is much more a stopping point, a reordering that prevents decoded semiotic flows in physics and mathematics from escaping in all directions. The axiomaticians are the men of the State of science, who seal off the lines of flight that are so frequent in mathematics, who would impose a new *nexum*, if only a temporary one, and who lay down the official policies of science.[36]

36. Gilles Deleuze and Felix Guattari, *A Thousand Plateaus*, trans. Brian Massumi, Minneapolis, University of Minnesota Press, 1987, p. 461.

What are your thoughts on the 'category' that Deleuze and Guattari propose here?

AB:

I would say that this text of Deleuze and Guattari is more focused on elaborating what an axiom is for them, rather than a real model in the sense of being connected to axioms. It is a very important distinction because I find that the very fundamental characteristic of formalistic thinking in general is axiomatic thought. The principal thing for me is that there are always axiomatic decisions. I do not believe at all that it is necessary to posit an opposition between the decision, on the one hand, and, on the other hand, intuition or experimentation, since the decision is always the formation or concentration of a series of experimentations or intuitions in a process of the perception of a situation. It is rather the notion of a common point between them that renders a network. Why? What is a decision? A decision is always a point where, at a given moment, there is something concentrated or crystallized where one faces a cut. It is a point that cuts and on this point, I would not say that axioms are on the side of state or official politics. This is not the case because it reduces the axiom to a formal making-explicit of things. Rather, we should say that the axiom is a regime of decision which is not at all identical with what is involved in being an 'axiomatician'. I do however agree that those who undertake the ordering of the axioms, people like Zermelo in set theory, or Peano for arithmetic, the task of organization that Bourbaki attempted, always arrive after the fact. Yet one should not be caught up with the idea that axioms are operative before all else. The decision is always taken in the very elements of the process itself so that perhaps the axiom is absolutely fundamental in every discipline of thought. It is creative, it itself is fundamentally creative. But one should not confound the explicit character of the axioms and the formalization of the axioms, which is a work of putting in order that takes place after the fact. It is like, in a certain way, a political organization that is really active, it exists before having its statutes, its declarations and all that; it exists in the process of its activities. As such, theoretically, I think I am situated quite differently. When I think of axiomatics, I think of the philosophy of Plato, where what is at stake are the principles and the axioms. I am opposed to a philosophy of definition. For me, then, the opposition is not between axioms and, on the other side, intuition and experimentation but rather the opposition between axioms and definition. Models are directly involved in this.

What does a model actually traverse? These are the very things through which we think the axioms

Perhaps here I would add a few things outside of the questions. I think that in rereading *The Concept of Model*, which is in the end what we are talking about, I would say something by way of a sort of an overview of the elements of continuity and discontinuity in my work .... After all, it has been 40 years. It is a morsel of existence (laughs). It is necessary always to begin with continuity because it is more important. The author of a work always likes to say that they have constantly evolved but I would like to pose the idea of Bergson that a philosopher only has one idea. If we suppose only one idea, it is this idea. I believe that if all creative thought is in reality the invention of a new mode of formalization, then that thought is the invention of a form. Thus if every creative thought is the invention of a new form, then it will also bring new possibilities of asking, in the end, 'what is a form?'[37] If this is true, then one should investigate the resources for this. As a resource, there is nothing deeper than that which the particularity of mathematics has to offer. This is what I think, I held this point of view and I hold it now. It is not that mathematics is the most important, not at all. Mathematics is very particular but in this, philosophically speaking, there is something that is specifically tied to mathematics in the very place of thought. Like Plato, who first thought this, thinking is the thinking of forms, something that he called ideas but they are also the forms. It is the same word, ἰδέα. It is different from Aristotle's thought where thinking is the thinking of substance. His paradigm is the animal. For Plato, it's mathematics. Mathematics holds something of the secret of thinking. It is that mathematics, while not the most important, is something which makes more transparent, or takes us closer to, this secret of thinking. This is the first point. I think I hold a fidelity to this idea, but, at the same time, the heart of the most radical experience is politics. Politics itself, in a sense, is also a thinking through forms. It is not the thought of arrangements or the thought of contracts

37. It should be noted that Badiou distinguishes philosophy and thinking in general. As such, the question of form is not primarily a question of philosophy but, for Badiou, a question of thinking and truth. Indeed, it is through the creation of forms, as stated here, that philosophy gains access into these various fields. At the heart of Badiou's philosophy, this distinction is central in his division of the four truth procedures, politics, art, love and science none of whom are interior to philosophy. Badiou's 1992 text *Conditions* is focused on this distinction. See Alain Badiou, *Conditions,* Paris, Éditions du Seuil, 1992, pp. 21-38, 103-118. His chapter on the distinction and relationship between philosophy and mathematics as well as his chapter on 'subtraction' has been translated into English in the collection *Theoretical Writings,* pp. 21-38, 103-118.

or the good life. No. It is a thinking of form. What is the radically new of the formalism of social relations and of community? This was the question for Plato. And finally there is a bizarre connection, a truly strange connection between politics and mathematics that has been totally central. It is not because there is a mathematics of politics, not at all. Rather the question concerns a politics that will allow for the secrets of thought. How does it understand thought marked out by mathematics? This is the question I would pose.

The differences, on the contrary, or the discontinuities, are that which concerns the question of the subject. At the time, when I wrote *The Concept of Model*, I thought that if mathematics were to achieve the secrets of thought it was because of its a-subjectivity. It seemed like a psychosis; that is to say, it was the automatism, a characteristic of the automatism of thought, a mechanical conception of mathematics that I was concerned with in those days. Thus the successive modifications are significant because I realized that that was not quite it. These very same elements of automatism were very close to treatments in the philosophy of language, formalism from the point of view of the 'linguistic turn'. More and more, I distanced myself from this. I began to think that if mathematics achieves the secrets of thought it was because of the type of thinking that it is. My conception of ontology began to follow this line of thought as well as the idea that the most sedimented thing will be pure multiplicity. I also began to think in terms of truth procedures. Finally, it is not at all surprising that it is most readable when its treatment is in a world the most saturated and complete.

I started by saying that I had a single idea. I have a very simple and minimal idea, that all thought is the opening up of a new type of formalization. Mathematics is only a particular kind of formalization among all the various differentiations and complex productions which stand in relation, on the one hand, to universality, the universality of form, and, on the other, to the particularity of its world.

TT:

Could it be that the question of discipline here relates both to the 'secrets' of mathematical thought and the discipline of the political field?[38]

---

38. In *Being and Event*, meditation twenty two, twenty three and twenty four, Badiou provides an overview of what he calls 'fidelity' or a 'discipline of time'. This discipline, in the context of the book is tied to various mathematical forms including the axiom of choice, forcing, generic

AB:

Absolutely. It is for this reason that the connection between discipline and fidelity are very important in the dimension that I call the ethics of my thought. I think that creative politics, the politics that changes the world, will not be at all the politics of spontaneity or of enthusiasm. In the end, it is a question of a politics of discipline, not a discipline of deduction but a discipline of practice.

TT:

Is there a Maoist theme here?

AB:

Yes, Maoist in a very deep sense. The discipline that Mao attempted, as we know now, is not best realized in the form of the party. I would say, if you like, that the party is like an out-moded mathematics... (laughs), that is to say, the mathematics of Euclid. We need to invent a non-Euclidian mathematics with respect to political discipline. Well, this is only an image. I think that Chairman Mao belonged to an epoch of the party: he was a Stalinist. But in the interior of this Stalinism, he realized this point of impasse. Having said this, he attempted to lay out another parameter for this discipline. As such, he plunged China into chaos, an utter chaos. But it was for profound reasons and not the reasons of a megalomaniac psychology, as they say today. The party is a Leninist realization of discipline, which is something absolutely interesting for the reason that it showed what would work at a certain point but remains in many ways insufficient. In being insufficient, it is absolutely necessary to invent a new political discipline of the process of the masses itself and not, in its place, an official 'axiomatics'. We should never be misled into thinking that we will not be able to find something to this end.

---

sets and the logic of deduction. In this sense, 'discipline' in mathematics has a certain echo in the idea of political discipline. *Being and Event*, p. 211.

Bibliography

Althusser, Louis, *Lire le Capital*, vol. 2, Paris, Presses Universitaires de France, 1965.

Althusser, Louis, *Eléments d'Autocritique*, Paris, Librairie Hachette, 1974.

Althusser, Louis, 'Elements of Self-Criticism', trans. Grahame Lock, in Lock (ed.), *Essays in Self-Criticism*, London, New Left Books, 1976.

Althusser, Louis, *Philosophy and the Spontaneous Philosophy of the Scientists*, Gregory Elliot (ed.), trans. Warren Montag, London, Verso, 1990.

Aoki, Douglas Sadao, 'Letters from Lacan', in *Paragraph*, vol. 29, no. 3, 2006, pp. 1-20.

Bachelard, Gaston, *L'activité de la physique rationaliste*, Paris, Presses universitaires de France, 1951

Bachelard, Gaston, *Le nouvel esprit scientifique*, 10th édition, Paris, Presses universitaires de France, 1968.

Bachelard, Gaston, *The Formation of the Scientific Mind*, trans. Mary MacAllester Jones, London, Clinamen Press, 2002.

Badiou, Alain, 'Le (Re)commencement du matérialisme dialectique', in *Critique*, vol. 23, no. 240, May 1967, pp. 438-367.

Badiou, Alain and Michel Serres, 'Modèle et Structure' (interview), *Philosophie* (Television programme), Jean Fléchet (dir.), Centre Nationale de Documentation Pédagogique, 1968.

Badiou, Alain, 'La subversion infinitesimale', in *Cahiers pour l'analyse*, vol. 9, 1968, pp. 118-137.

Badiou, Alain, *Le Concept de modèle*, Paris, François Maspero, 1969.

Badiou, Alain, 'Marque et Manque: a propos de zéro', *Cahiers pour l'analyse*, no. 10, 1969, pp. 150-173.

Badiou, Alain, *Théorie de la contradiction*, Paris, Maspero, 1975.

Badiou, Alain, *De l'idéologie*, Paris, Maspero, 1976.

Badiou, Alain, *Theorie du Sujet*, Paris, Seuil, 1982.

Badiou, Alain, 'Orientation de pensée transcendante', seminar on October 24th, 1987 at Paris University VII. Notes transcribed by François Nicolas.

Badiou, Alain, *L'être et l'événement*, Paris, Éditions de Seuil, 1988.

Badiou, Alain, 'Saissement, dessaisie, fidélité', in *Les Temps modernes*, vol. 531, 1990, pp. 14-22.

Badiou, Alain, *Le Nombre et les nombres*, Paris, Éditions du Seuil, 1990.

Badiou, Alain, 'On a Finally Objectless Subject', trans. Bruce Fink, in Eduardo Cadava et al. (eds.), *Who Comes After the Subject?* London, Routledge, 1991, pp. 24-32.

Badiou, Alain, *Conditions*, Paris, Éditions du Seuil, 1992.

Badiou, Alain, *Manifesto for Philosophy*, trans. Norman Madarasz, Albany, SUNY Press, 1999.

Badiou, Alain, *Ethics: An Essay on the Understanding of Evil*, trans. Peter Hallward, London, Verso, 2001.

Badiou, Alain, *Saint Paul and the Foundations of Universalism*, trans. Ray Brassier, Stanford, Stanford University Press, 2003.

Badiou, Alain, *Theoretical Writings*, trans. and ed. Ray Brassier and Alberto Toscano, London, Continuum, 2004.

Badiou, Alain, *Metapolitics*, trans. Jason Barker, London and New York, Verso, 2005.

Badiou, Alain, *Being and Event*, trans. Oliver Feltham, London, Continuum, 2005.

Badiou, Alain, 'Can Change be Thought?', with Bruno Bosteels, in Gabriel Riera (ed.), *Alain Badiou: Philosophy and its Conditions*, Albany, SUNY Press, 2005, pp. 237-261.

Badiou, Alain, *Briefings on Existence: A short treatise on transitory ontology*, trans. Norman Madarasz, Albany, SUNY Press, 2006.

Badiou, Alain, *Logiques des mondes*, Paris, Éditions du Seuil, 2006.

Badiou, Alain, *The Century*, trans. Alberto Toscano, Cambridge, Polity, 2007.

Badiou, Alain, 'New Horizons in Mathematics as a Philosophical Condition: An Interview with Alain Badiou', with Tzuchien Tho, *Parrhesia*, no. 3, 2007, pp. 1-11.

Badiou, Alain, 'Préface de la nouvelle édition', *Le Concept de modèle*, Paris, Fayard, 2007.

Badiou, Alain, *Le Concept de modèle*, Paris, Fayard, 2007.

Badiou, Alain, 'Infinitesimal Subversion', trans. Robin Mackay, in Peter

Hallward and Christian Kerslake (eds.), *Concept and Form: The Cahiers pour l'analyse and Contemporary French Thought*, , forthcoming, 2010.

Badiou, Alain, 'Mark and Lack', trans. Zachary Fraser, in Peter Hallward and Christian Kerslake (eds.), *Concept and Form: The Cahiers pour l'analyse and Contemporary French Thought*, , forthcoming, 2010.

Balibar, Etienne, *Cours de philosophie pour scientifiques*, [Unpublished].

Bernays, Paul, 'What do some recent results in set theory suggest?', in Imre Lakatos (ed.), *Problems in the Philosophy of Mathematics*, Amsterdam, North Holland Publishing Company, 1967.

Bosteels, Bruno, 'Alain Badiou's Theory of the Subject: The Recommencement of Dialectical Materialism? Part I', *Pli: Warwick Journal of Philosophy*, vol. 12, 2001, pp. 200-229.

Bosteels, Bruno, 'Alain Badiou's Theory of the Subject: The Recommencement of Dialectical Materialism? Part II', *Pli: Warwick Journal of Philosophy*, vol. 13, 2002, pp. 173-208.

Brassier, Ray, 'On Badiou's Materialist Epistemology of Mathematics', in *Angelaki*, vol. 10, n.2, 2005, pp. 135-150.

Blanché, R. *Axiomatics*, trans. G. B. Keene, London, Routledge & Kegan Paul, 1962.

Bourbaki, Nicholas, *Théorie des ensembles*, Paris, Hermann, 1968.

Bourbaki, Nicholas, *Theory of Sets*, New York, Springer, 2004.

Butler, Judith, Ernesto Laclau and Slavoj Žižek, *Contingency, Hegemony, Universality*, London, Verso, 2000.

Canguilhem, Georges, 'L'expérimentation en biologie animale', in *La connaissance de la vie*, 2e édition, Paris, J. Vrin, 1965.

Cantor, Georg, 'Letter to Dedekind', in Jean van Heijenhoort (ed.), *From Frege to Gödel: A Source Book in Mathematical Logic, 1879-1931*, Cambridge, Harvard University Press, 1967, pp. 113-117.

Carnap, Rudolf, 'Logical Foundations of the Unity of Science', in Otto Neurath et al. (eds.), *International Encyclopedia of Unified Science*, vol. 1, Chicago, University of Chicago Press, 1955.

Clemens, Justin, 'Letters as the Condition of Conditions for Alain Badiou', in *Communication & Cognition*, vol. 36, no. 1-2, 2003, pp. 73-102.

Cohen, Paul J., *Set Theory and the Continuum Hypothesis*, W. A. Benjamin, 1966.

Cundy H. M. & A. P. Rollet, *Mathematical Models*, Oxford, Clarendon Press, 1961.

Curry, Haskell B. *Leçons de logique algébrique*, Paris, Gauthier-Villars, 1952.

Deleuze, Gilles and Felix Guattari, *A Thousand Plateaus*, trans. Brian Massumi, Minneapolis, University of Minnesota Press, 1987.

Descartes, René, *Meditations on First Philosophy*, trans. John Veitch, London, Everyman's Library, 1969.

Euclid, *Elements*, trans. Sir Thomas Heath, in Robert Maynard Hutchins (ed.), *Great Books of the Western World*, vol. 2, London, William Benton Press, 1952.

Fraser, Zachary Luke, 'The Law of the Subject: Alain Badiou, Luitzen Brouwer and the Kripkean Analyses of Forcing and the Heyting Calculus', in Paul Ashton, A. J. Bartlett & Justin Clemens (eds.), *The Praxis of Alain Badiou*, Melbourne, re.press, 2006, pp. 23-70.

Frege, Gottlob, *Foundations of Arithmetic: A Logico-Mathematical Inquiry into the Concept of Number*, trans. J. L. Austin, New York, Harper, 1960.

Hallward, Peter, *Badiou: A Subject to Truth*, Minneapolis, University of Minnesota Press, 2003.

Hilbert, David, *The Foundations of Geometry*, trans. E. J. Townsend, La Salle, Illinois, Open Court, 1965.

Hilbert, David, 'The Foundations of Mathematics', in Jean van Heijenhoort (ed.), *From Frege to Gödel: A Source Book in Mathematical Logic, 1879-1931*, Cambridge, Harvard University Press, 1967, pp. 464-479.

Hilbert, David, 'Letter to Frege dated December 29th, 1899' trans. Hans Kaal, in Gottlob Frege, *Philosophical and Mathematical Correspondence*, Gottfried Gabriel (ed.), Chicago, University of Chicago Press, 1980, pp. 38-42.

Hilbert, David and S. Cohn-Vossen, *Geometry and the Imagination*, trans. P. Nemenyi, New York, Chelsea Pub. Co., 1952.

Kant, Immanuel, *Critique of Pure Reason*, trans. Norman Kemp Smith, New York, Palgrave Macmillan, 2003.

Kreisel, G. and J.L. Krivine, *Elements of Mathematical Logic (Model Theory)*, Amsterdam, North-Holland Publishing Company, 1967.

Krivine, Jean-Louis, *Théorie axiomatique des ensembles,* Paris, Presses Universitaires de France, 1969.

Lacan, Jacques, *Écrits*, Paris, Éditions du Seuil, 1966.

Lacan, Jacques, 'L'Instance de la lettre dans l'inconscient', *Écrits*, Paris, Éditions du Seuil, 1966, pp. 493-528.

Lacan, Jacques, 'La séminaire sur "La Lettre volée"', *Écrits*, Paris, Éditions du Seuil, 1966, pp. 11-60.

Lacan, Jacques, *Écrits*, trans. Bruce Fink, New York, W. W. Norton, 2006.

Lacan, Jacques, 'The Mirror Stage as Formative of the *I* Function as Revealed in Psychoanalytic Experience', in *Écrits*, trans. Bruce Fink, New York, W. W. Norton, 2006, pp. 75-81.

Lautman, Albert, *Les mathématiques, les idées et le réel physique*, Paris, Libraire Philosophique J. Vrin, 2006.

Lawvere, F. W. and Robert Rosebrugh, *Sets for Mathematics*, New York, Cambridge University Press, 2003.

Lenin, V.I. *Materialism and Empirio-criticism*, trans. Anon., Peking, Foreign Languages Press, 1972.

Lévi-Strauss, Claude. 'Social Structure', in *Structural Anthropology*, vol. 1, trans. Claire Jacobson & Brooke Grundfest Schoepf, London, Basic Books, 1963.

Martin, Roger, *Logique contemporaine et formalisation*, Paris, Presses universitaires de France, 1964.

Marx, Karl, 'Theses on Feuerbach', in Robert C. Tucker (ed.), *The Marx-Engels Reader*, New York, W. W. Norton, 1978, pp. 143-5.

Mendelson, Elliott, *Introduction to Mathematical Logic*, Princeton, Van Nostrand, 1964.

Miller, Jacques-Alain, 'L'action de la structure', *Cahiers pour l'analyse*, no. 9, 1968.

Miller, Jacques-Alain, 'La suture (elements de la logique du signifiant)', *Cahiers pour l'analyse*, no. 1, 1966.

Pearson, Karl, *The Grammar of Science*, London, J. M. Dent & Sons, 1899.

Pêcheux, Michel and Michel Fichant, *Sur l'histoire des sciences*, Paris, F. Maspero, 1969.

Poitou, Georges, *Introduction à la théorie des catégories*, Paris, Offilib, 1967.

Quine, W. V. O., 'On What There Is', in *From a Logical Point of View: Nine Logico-Philosophical Essays*, Cambridge, Harvard University Press, 2006.

Quine, W. V. O. *From a Logical Point of View: Nine Logico-Philosophical Essays*, Cambridge, Harvard University Press, 2006.

De Saussure, Ferdinand, *Course in General Linguistics*, trans. Wade Baskin, Toronto, McGraw-Hill, 1959.

Schotch, Peter, *Introduction to Logic and its Philosophy*, electronic resource, 2006. Available at http://www.schotch.ca.

Serres, Michel, *Le système de Leibniz et ses modèles mathématiques*, 2 vols., Paris, Presses Universitaires de France, 1968.

Tiles, Mary, *Mathematics and the Image of Reason*, London, Routledge, 1991.

Von Neumann, John and Oskar Morgenstern, *Theory of Games and Economic Behaviour*, Princeton, Princeton University Press, 1953.

Winterbourne, A. T., *The Ideal and the Real*, London, Kluwer, 1988.

Wittgenstein, Ludwig, *Tractatus Logico-Philosophicus*, trans. C. K. Ogden, London, Routledge, 1922.

Žižek, Slavoj, '*Da capo senza fine*', in Judith Butler, Ernesto Laclau & Žižek, Slavoj, *Contingency, Hegemony, Universality*, London, Verso, 2000, pp. 213-252.

Žižek, Slavoj, *The Ticklish Subject*, London, Verso, 2000.

About the Authors

Alain Badiou was born in Rabat, Morocco in 1937. He studied at the Ecole Normale Supérieure in the 1950s and, from 1969 until 1999, taught at the University of Paris VIII (Vincennes-Saint Denis) before returning to ENS as the Chair of the philosophy department. Much of Badiou's life has been shaped by his dedication to the consequences of the May 1968 revolt in Paris. Long a leading member of Union des jeunesses communistes de France (marxistes-léninistes), he remains, with Sylvain Lazarus and Natacha Michel, at the centre of L'Organisation Politique, a post-party organization concerned with direct popular intervention in a wide range of issues, including immigration, labor, and housing. Alain Badiou is the author of several successful novels and plays, as well as more than a dozen philosophical works.

Zachary Luke Fraser teaches in the Contemporary Studies and History of Science and Technology Programmes at the University of King's College in Halifax, Nova Scotia, where he lives with his wife, Shan Mackenzie-Fraser, and their four children, two dogs and two cats. He has written extensively on Alain Badiou and Jean-Paul Sartre, and is the author of "The Law of the Subject" in *The Praxis of Alain Badiou*, also published by re.press. His current research is taking place through the Philosophy and the History of Science and Technology departments at Dalhousie University and King's College, where he is working towards a logical formalization of the archaeology of knowledge.

Tzuchien Tho is a graduate student of Philosophy at the University of Georgia (USA) and University of Paris X. He is currently writing a dissertation on Leibniz and working on questions surrounding infinitesimals, Platonism and early modern mathematics and metaphysics.

Printed in the United States
110257LV00001B/58/A